SpringerBriefs in Energy

More information about this series at http://www.springer.com/series/8903

Sushant Kumar

Clean Hydrogen
Production Methods

 Springer

Sushant Kumar
Department of Mechanical
  and Materials Engineering
Florida International University
Miami, FL
USA

ISSN 2191-5520   ISSN 2191-5539   (electronic)
SpringerBriefs in Energy
ISBN 978-3-319-14086-5   ISBN 978-3-319-14087-2   (eBook)
DOI 10.1007/978-3-319-14087-2

Library of Congress Control Number: 2014957723

Springer Cham Heidelberg New York Dordrecht London

Printed on acid-free paper

Springer International Publishing AG Switzerland is part of Springer Science+Business Media
(www.springer.com)

In loving memory of my father

Foreword

The book discusses solutions to the existing problem related to the production of hydrogen. Similar to electricity, hydrogen is a high-efficiency energy carrier, which can lead to zero or near-zero emissions at the point of use. Ironically, the production of hydrogen is an energy intensive process and thus nullifies its inherent advantage. The required energy is mainly provided by burning fossil fuel which subsequently emits a huge amount of CO_2 per unit of hydrogen. Over the years, a lot of attention has been given to mitigate emission of CO_2 during hydrogen production. This book presents various clean methods to produce hydrogen using both fossil and non-fossil energy sources. Moreover, the book illustrates novel techniques based on the inclusion of oxides and hydroxides in the existing major hydrogen production technologies. Such methods provide a local minimal solution to a global problem. Depending on how the carbon binding solid is used, the carbon dioxide emission may be delayed or permanently prevented.

The information in this book can be useful for engineers, scientists, researchers and practitioners involved in the various fields of hydrogen production and CO_2 mitigation techniques. The book reviews the latest development in these areas, presents novel methods and also critically evaluates the efforts made so far.

Miami, FL, USA

Prof. S.K. Saxena
CeSMEC, Florida International University

Acknowledgments

I would like to express my thanks and gratitude to Prof. S.K. Saxena for his invaluable guidance, encouragement, excellent comments throughout this project. His constructive criticism and critical review were absolute blessings for my book. I offer my heartfelt thanks to Dr. Vadym Drozd for his continuous suggestions and help during the start of this research work. Also, I would like to thank Elizabeth Hawkins and Judith Hinterberg for helping me to write for the highly prestigious SpringerBriefs in Energy. I am also thankful to my friends, Nikoloz Esitashvili and Vivek Rao, who supported me to achieve this work of art. Moreover, I shall be grateful to University Graduate School, FIU for supporting me with through Dissertation Evidence Acquisition (DEA) and Dissertation Year Fellowship (DYF) awards.

I fall short of words to express my wholehearted gratitude to my family for being my source of energy. Their blessings, selfless and unfeigned love and empathy always motivated me to be happy. I would like to thank my mother for her eternal love and sacrifice for me. This book is a gift to my mother. Last but not the least, thanks to Jessica Bartley.

Contents

Chapter 1
Role of Hydrogen in the Energy Sector

Abstract The vast depletion of fossil fuels, the increase in carbon dioxide levels in the atmosphere, and the related environmental hazards represent a growing concern for the mankind. Therefore, over the past few decades, significant efforts have been made to establish hydrogen economy. Hydrogen is a high-efficiency energy carrier, which can lead to zero or near-zero emissions at the point of use. Moreover, it has been technically shown that hydrogen can be used for transportation, heating, and power generation, and could replace current fuels in all the present applications. Besides the challenge of storing hydrogen, development of clean hydrogen production methods is considered as a prime hindrance to establish the hydrogen economy. Here, the focus is to provide a brief overview of all the processes based on both renewable and non-renewable energy sources that have been proposed to produce clean hydrogen.

Keywords Fossil fuels · Hydrogen economy · Clean hydrogen

1.1 Introduction

Fossil fuels, which include oil, natural gas, and coal, continue to be the primary energy source for electricity, transportation, and residential services. Formed from organic material over the course of millions of years, fossil fuels have significantly contributed to global development over the past century. According to recent figures published by the US Department of Energy and Energy Information Administration, the world energy consumption is projected to increase from 524 quadrillion British thermal unit (BTU) [1 quadrillion BTU = 1.055×10^{18} J] in 2010 to as high as 820 quadrillion BTU by 2040 (Energy Information Administration, 2013) [1]. In other words, the world energy demand will grow by 56 % between 2010 and 2040. Owing to the economic growth and expanding population, the global energy consumption is mostly concentrated in the developing countries [non-organization for Economic Cooperation and Development (OECD)]. As can be seen in Fig. 1.1a, the use of

© The Author(s) 2015   1
S. Kumar, *Clean Hydrogen Production Methods*,
SpringerBriefs in Energy, DOI 10.1007/978-3-319-14087-2_1

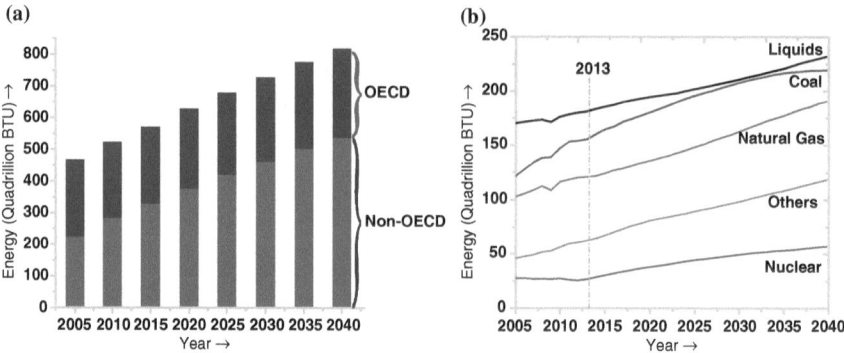

Fig. 1.1 World energy consumption by **a** nations **b** fuels [1]

energy will grow by 90 % in non-OECD countries; in OECD countries, the increase is only 17 % [1].

Although renewable energy and nuclear power are the world's fastest growing energy sources, fossil fuels are expected to supply almost 80 % of world energy use through 2040. Natural gas is the fastest growing fossil fuel, increasing by 1.7 % per year. The growth in the contribution of natural gas is due to the increasing supply of tight gas, shale gas, and coal bed methane. The use of coal is assumed to grow faster than petroleum and other liquid fuel till 2030, primarily because of its growing demand in the developing countries (Fig. 1.1b).

Fossil fuels are attractive because of their relatively low cost compared with renewable sources such as biomass, wind, geothermal, and solar power. As fossil fuels have limited reserves, the crucial question is "how long these resources will last." According to Edigera et al., the fossil fuel production for Turkey will be exhausted by 2038 [2, 3]. Similarly, India, China, Russia, and USA will see their coal reserves depleted in 315, 83, 1,034, and 305 years, respectively [4]. Countries such as the USA, India, and Ethiopia are already looking for new coal reserves and investing to innovate clean coal technologies [5, 6]. On the contrary, Lior et al. argued that despite the rise in consumption, more reserves are found or exploited, leading to a higher amount of fossil fuels available on the world market, and the ratio of resources to production has been nearly constant for decades, around 40, 60, and 150 for oil, gas, and coal, respectively [7]. In 2009, Shafiee et al. modified the Klass model (the model calculates the ratio of reserves usage and helps to predict the depreciation time of fossil fuels [8]) and assumed a continuous compound rate. The work predicts that oil and gas will be exhausted earlier than coal [9]. The depletion time for oil, gas, and coal is computed to be around 35, 37, and 107 years, respectively. Therefore, coal reserves will most likely be the only fossil fuel remaining after 2042 and will be exhausted by 2112.

There are so many factors affecting future projections that it can be difficult to prepare any model that can precisely show the future trends in natural resources. The factors affecting future predictions for fossil fuels include economic growth, production and consumption rate, total reserves, development of unconventional methods for extracting resources, and population growth. Despite the differences in time estimates, it is certain that these resources can only continue for a finite period.

Another major concern in using fossil fuels is the emission of greenhouse gases such as carbon dioxide (CO_2). Natural greenhouse gases (CO_2, CH_4, N_2O, CFCs, HFCs, PFCs, SF_6, and H_2O) absorb a considerable fraction of solar thermal radiation. Further, these gases reradiate the solar energy to the surface of the earth in the form of visible light and thus prevent heat from escaping the earth's atmosphere. Trapping this heat in the atmosphere raises the temperature of the earth to 33 °C, making it habitable [10]. However, concentration of greenhouse gases in the atmosphere has increased significantly since the industrial revolution [11]. CO_2 is considered as the prime contributor to global warming and accounts for 64 % of the increased greenhouse effect [12]. Recent studies do indicate that global warming is human induced [13]. There is a growing belief that if such extensive use of fossil fuels continues for another 50 years, the CO_2 concentration will rise to 580 ppm, which would trigger a severe climate change [14].

In early 2014, the atmospheric CO_2 concentration was 397.8 ppm, which is about 42 % higher compared to that in year 1800 [1]. Figure 1.2a depicts the world energy-related CO_2 emissions. According to International Energy Outlook (IEO) 2013, the emission is projected to increase from 31.2 billion metric tons in 2010 to 45.5 billion metric tons in 2040 [1]. Non-OECD countries rely on the supply of fossil fuels to fulfill their energy needs and consequently emits huge amount of CO_2. Figure 1.2b illustrates the annual CO_2 emissions form the use of different fuel types such as coal, liquid fuels, and natural gas. Among fossil fuels, the use of coal itself contributed to 44 % of overall CO_2 emission in 2010 and is projected to increase to 47 % in 2020–2030, before dropping marginally to 45 % in 2040. On other hand, liquid fuels have the slowest growth, resulting in an increase of only 3.5 billion metric tons of CO_2 from 2010 to 2040. As can be seen from Fig. 1.2b, the consumption of natural gas is growing faster compared to coal or liquid fuel.

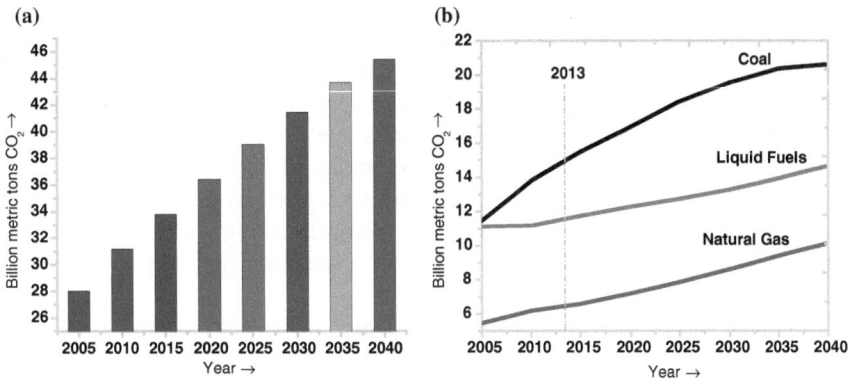

Fig. 1.2 World energy-related carbon dioxide emissions by **a** year **b** fuel type [1]

As natural gas has relatively low carbon intensity, its contribution to world energy-related CO_2 emissions might be only 22 % in 2040.

The current world energy requirement for transportation and heating (2/3rd of the primary energy demand) is mainly supplied by petroleum and natural gas. These two fuels are preferred due to the relative ease of transportation of liquid or gaseous forms. It is noteworthy that the combustion of hydrocarbon fuels for transportation and heating contributes over half of all greenhouse gas emissions and a large fraction of air pollutant emissions [15]. It has been estimated that the carbon emissions have to be reduced to one-third of its present value in order to maintain the CO_2 level in the atmosphere while meeting the ever-increasing global energy demands.

In essence, the vast depletion of fossil fuels, the increase in carbon dioxide levels in the atmosphere, and the related environmental hazards represent a growing concern for the mankind. The enormity of global warming is daunting. Thus, one of the proposed solutions is to use an alternative fuel.

1.2 Hydrogen as an Alternative Fuel

An ideal alternative fuel should be inexpensive, convenient to use, clean, and have lower carbon content. Among various alternatives, hydrogen fuel offers the highest potential benefits and possesses most of the key criteria for an ideal fuel. Similar to electricity, hydrogen is a high-efficiency energy carrier, which can lead to zero or near-zero emissions at the point of use. Therefore, for past 50 years, researchers and several industrial organizations have promoted hydrogen fuel as the solution to the global warming. However, hydrogen economy is not a new concept. In 1874, Jules Verne stated that "water will be the coal of the future" [16]. After few decades, Rudolf Erren suggested the use of hydrogen as a transportation fuel [17].

Hydrogen is a potentially emissions-free alternative fuel with a very high specific energy content of about 140.4 MJ/kg (gasoline has only 48.6 MJ/kg) [15]. The huge interest for hydrogen production and utilization is based on the premise that the fuel cell is a proven technology and hydrogen is abundant on the Earth. However, hydrogen on the Earth is in its oxidized state (H_2O), which has no fuel value, and there are no other natural resources for hydrogen. Fortunately, hydrogen can be produced using both renewable and non-renewable resources. The available technologies for hydrogen production are the reforming of natural gas, gasification of coal and biomass, and the splitting of water by water electrolysis, photo-electrolysis, photobiological production, water-splitting thermochemical cycle, and high temperature decomposition. Figure 1.3 illustrates the sources and methods of hydrogen production [18].

Currently, steam methane reformation (SMR) is the most common and least expensive method to produce hydrogen. It is a two-step process. At first step, methane reacts with steam at temperature 700–1100 °C to form Syn gas ($CO + H_2$), and then, carbon monoxide reacts with steam to produce additional hydrogen.

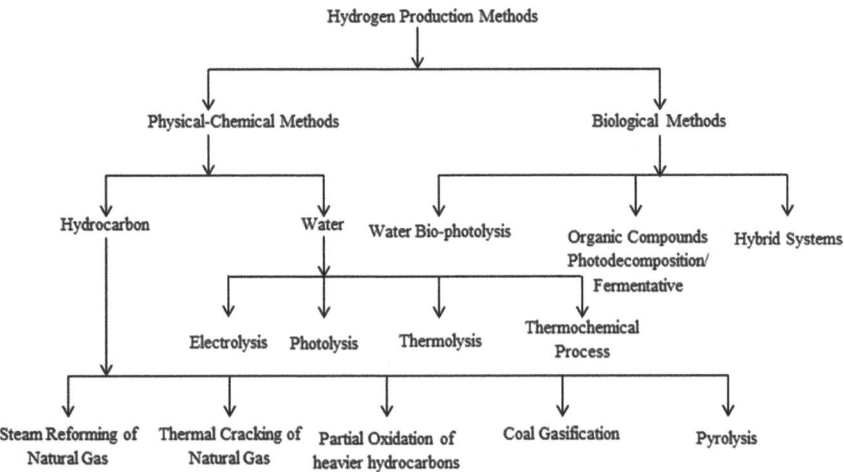

Fig. 1.3 Sources and methods of hydrogen production

Hydrogen, when produced from reforming of hydrocarbons, generates CO_2 as a by-product. About 2.5 t of CO_2 is vented into the atmosphere for each ton of hydrogen produced from reforming of hydrocarbons [19, 20]. Similarly, coal gasification process is another such matured technology and has no exception to huge CO_2 emissions. Here, coal (CH_xO_y) reacts with O_2 or steam to produce hydrogen. Such enormous emission of CO_2 downgrades the use of the conventional techniques to produce hydrogen. Thus, it is essential to develop methods that can produce hydrogen without or with reduced CO_2 emissions. A brief overview of the proposed CO_2 capture techniques for the SMR and coal gasification processes is provided in Chaps. 3 and 4, respectively.

A high CO_2 emission during hydrogen production nullifies its inherent advantages. Thus, the focus is to review the modified or new hydrogen producing technologies that do not lead to carbon emissions. Another possibility is to capture the CO_2 produced in these processes to avoid its release to the atmosphere. There are mainly two such ways to capture CO_2: (i) in situ capture (at the point of emission) and (ii) from the air (direct capture). However, at present, the input energy for CO_2 capture processes is mainly provided by burning fossil fuels. Thus, it is obvious that to run such CO_2 capture processes may accelerate the depletion rate of the global fossil fuel reserves. Several methods have been proposed to mitigate overall carbon emission during hydrogen production, but most of them are either expensive compared to those using fossil fuels, or are in their very early stages of development.

Interestingly, hydrogen can also be produced using non-fossil resources, primarily water [21, 22]. Unlike hydrocarbons, water does not emit $CO_2 [H_2O \rightarrow H_2 + 1/2O_2]$ directly during hydrogen production. However, the direct splitting of water is a very energy intensive process and generally requires very high temperature (>2,000 °C) [23]. Of course, generation of such high temperature involves the burning of fossil fuels. Various techniques have been suggested to directly split water and that includes electrolysis, photoelectrochemical, photocatalytic, photobiological, and thermal decomposition. Among these, electrolysis process has potential to be a viable process at large scale in the midterm future. Moreover, the efficiency of the electrolysis of water is favorable (~ 75 %), but the cost of generation is several times higher than that from fossil fuels [24, 25]. The high endothermic water-splitting process, when assisted by burning fossil fuels, leads to huge CO_2 emissions. Thus, renewable energy sources such as solar or wind need to be developed for the electrolysis of water in the foreseeable future.

Fortunately, solar energy can also be used to break water into its chemical constituents (hydrogen and oxygen). Recently, US Department of Energy also suggested that the solar photodecomposition of water can be the long-term CO_2-free method for mass production of hydrogen [26]. However, the solar-based process requires huge land area, which is a matter of concern. In this regard, several modifications to solar-based process have been recently proposed. For instance, hydrogen can also be produced from metal/metal oxide systems with the aid of solar energy [27]. These reactions generally involve two steps: (1) dissociation of metal oxide to metal and (2) reaction of metal with steam to produce hydrogen. Such reactions are thermodynamically more favorable than the direct water-splitting process. However, one more challenge is to develop construction material that can withstand such high temperature, thermal shock, and oxygen, as well as prevent the reverse reaction in the first step while maintaining high efficiency. The investigation shows that the inclusion of H_2SO_4 or HI in thermal chemical cycles can reduce the operating temperature to approximately 850 °C or approximately 450 °C, respectively [28]. Recently, Ibrahim et al. summarized the solar-based hydrogen production systems and can be found elsewhere [29].

In addition, the next generation fission reactors can utilize heat to directly split water. Hydrogen production using nuclear energy offers one of the most attractive strategies and can be considered as a potential candidate for large-scale hydrogen source. The net efficiency of the process is the product of the efficiency in the reactor in producing electricity, times the efficiency of the electrolysis cell. Thermochemical water-splitting process has heat-to-hydrogen efficiencies of 50 % and thus gained substantial attention [30]. Nuclear energy can generate hydrogen in several ways which includes: (a) electrolysis of water, (b) high temperature electrolysis using minor heat and major electricity from the nuclear reactor, (c) nuclear heated steam reforming of natural gas, and (d) thermochemical water-splitting process using major heat and minor electricity from the nuclear reactor [15]. Currently, the fission process cannot be considered as a sustainable alternative. However, a demonstration of safe recycling of nuclear fuel may gain the social acceptance for nuclear fission process and can pave its pathway for large-scale application.

In the places where natural gas is scarce, on-site hydrogen can be produced from water, methanol, or ammonia, via electricity derived from renewable energy sources such as solar, wind, or biofuels [31, 32]. Wind energy is abundant, clean, safe, and inexhaustible source of energy with only a minor environmental impact caused mostly during the time of installation of its equipment [33]. As a result, the efforts have been made for the electrolytic hydrogen production using large-scale wind installations. However, wind energy system cannot produce electrical energy at a very high percentage in a system, and thus, its installation can be advantageous only in certain sites with high wind potential and suitable geographical conditions. The electricity generated from wind energy can be coupled with the hydrogen producing stacks of electrolyzers and that would reduce energy conversion losses and capital costs investment. The cost of hydrogen production is estimated to be $4/kg or less at high wind class sites, class 4, or higher [34].

Interestingly, solar and wind power do not need water to generate electricity. So these resources are not only sustainable carbon-free system, but also reduce water requirements for electricity generation. Photobiological [35] and Photo-electrochemical processes [36–38] are solar- and water-driven methods for the economic production of hydrogen. It is obvious that the electricity derived from renewable sources such as wind and solar may generate local hydrogen, but certainly will not meet up the global hydrogen demand. Similar limitation exists with the application of biological reformation of biomass using microorganisms. Although biomass is a carbon neutral fuel, use of fertilizers for its production incurs "CO_2 cost." As the synthesis of fertilizers requires ammonia, which itself is produced from hydrogen and nitrogen, the former mainly produced by burning of fossil fuels. Additionally, biomass can be converted to valuable liquid fuels such as methanol, ethanol, biodiesel, and pyrolysis oil, which could be transported and used to produce hydrogen on-site. Since biomass can either act as a chemical feedstock or an energy source, the obvious question arises: how efficiently it should be used in this food and carbon constrained world? [39].

In summary, continuous process and equipment improvements made over the past 25 years have led to a reduction of 20 % of CO_2 emissions at hydrogen plants [40]. However, still a lot needs to be done to further reduce the CO_2 emission during hydrogen production. It seems certain that meeting the global energy demand, requires the exploration and deployment of a clean fossil fuels-based hydrogen production method in the coming years.

1.3 Conclusion

As discussed in this chapter, the continuous use of fossil fuels for hydrogen production faces severe challenges. Besides hydrogen storage, clean hydrogen production is seen as another big obstacle to realize hydrogen economy. Most of the conventional hydrogen production techniques emit huge amount of CO_2 per ton of hydrogen generated. Such vast CO_2 emission nullifies the inherent advantage of hydrogen. Thus, the modification of existing conventional methods or development of innovative methods is necessary. Fortunately, in recent times, several methods have been investigated and developed that can resolve the CO_2 emission issue during hydrogen production. Some of them use renewable energy sources, while others are merely modified version of existing fossil fuel-based methods. Considering the fact that renewable sources may fall behind meeting up the global hydrogen demand, and here, considerable attention is given to technologies based on non-renewable energy sources. A detailed information about the new and modified fossil fuel-based clean hydrogen production techniques are provided in the following chapters.

References

1. US Energy Information Administration (EIA), International Energy Statistics database (as of Nov. 2013) www.eia.gov/ies.Projections: EIA, Annual Energy Outlook 2014, DOE/EIA-0383 (214) (Washington, DC: April 2014) AEO 2014. National Energy Modeling System, run REF 2014.D1024BA www.eia.gov/aeo
2. Edigera VS, Akar S, Ugurlu B (2006) Forecasting production of fossil fuel sources in Turkey using a comparative regression and ARIMA model. Energy Policy 34:3836–3846. doi:10. 1016/j.enpol.2005.08.023
3. Edigera VS, Akar S (2007) ARIMA forecasting of primary energy demand by fuel in Turkey. Energy Policy 35:1701–1708. doi:10.1016/j.enpol.2006.05.009
4. Asif M, Muneer T (2007) Energy supply, its demand and security issues for developed and emerging economies. Renew Sustain Energy Rev 11:1388–1413. doi:10.1016/j.rser.2005.12. 004
5. Khadse A, Qayyumi M, Mahajani S, Aghalayam P (2007) Underground coal gasification: a new clean coal utilization technique for India. Energy 32:2061–2071. doi:10.1016/j.energy. 2007.04.012
6. Wolela A (2007) Fossil fuel energy resources of Ethiopia: coal deposits. Int J Coal Geol 72:293–314. doi:10.1016/j.coal.2007.02.006
7. Lior N (2008) Energy resources and use the present situation and possible paths to the future. Energy 33:842–857. doi:10.1016/j.energy.2009.06.049
8. Muda N, Pin TJ (2012) On prediction of depreciation time of fossil fuel in Malaysia. J Math Stat 8:136–143. doi:10.3844/jmssp.2012.136.143
9. Shafiee S, Topal E (2009) When will fossil fuel reserves be diminished? Energy Policy 37:181–189. doi:10.1016/j.enpol.2008.08.016
10. Iyer M (2006) High temperature reactive separation process for combined carbon dioxide and sulfur dioxide capture from flue gas and enhanced hydrogen production with in-situ carbon dioxide capture using high reactivity calcium and biomineral sorbents. Electronic thesis or Dissertation Ohio State University https://etd.ohiolink.edu/
11. Caldeira K (2006) Forests, climate, and silicate rock weathering. J Geochem Explor 88:419–422. doi:10.1016/j.gexplo.2005.08.089
12. Bryant E (1997) Climate process and change. Cambridge University Press, Cambridge, p 118
13. Oreskes N (2004) The scientific consensus on climate change. Science 306:1686. doi:10.1126/science.1103618
14. Fan LS (2010) Chemical looping systems for fossil energy conversions. Wiley, Hoboken, New Jersey Chapter 1, p 12
15. Gupta R (2008) Hydrogen fuel: production, transport and storage. CRC Press, FL. Chapter 1, p 9
16. Verne J (1874) The mysterious island. Available at http://www.literature-web.net/verne/mysteriousisland
17. Hoffmann P (1981) The forever fuel: the story of hydrogen. Westview Press, Boulder, CO. Chapter 6, p 164

18. Krishna RH (2013) Review of research on production methods of hydrogen: future fuel. Eur J Biotechnol Biosci 1:84–93. (http://www.biotechjournal.info/vol1/issue2/pdf/25.1.pdf)
19. EIA (2025) Annual energy outlook with projections to 2025. Washington
20. Blazek CF, Biederman RT, Foh SE, Jasionowski W(1992) Underground storage and transmission of hydrogen. In: Proceedings of the third annual us hydrogen meeting. Washington, March 18–20, pp 4–221
21. Sherif SA, Barbir F, Veziroglu TN (2005) Wind energy and the hydrogen economy—review of the technology. Sol Energy 78:647–660. doi:10.1016/j.solener.2005.01.002
22. Penner SS (2006) Steps towards the hydrogen economy. Energy 31:33–43. doi:10.1016/j.energy.2004.04.060
23. Funk JE (2001) Thermochemical hydrogen production: past and present. Int J Hydrogen Energy 26:185–190. doi:10.1016/S0360-3199(00)00062-8
24. Ewan BCR, Allen RWK (2005) A figure of merit assessment of the routes to hydrogen. Int J Hydrogen Energy 30:809–819. doi:10.1016/j.ijhydene.2005.02.003
25. Edwards PP, Kuznetsov VL, David WIF (2007) Hydrogen Energy. Phil Trans R Soc A 365:1043–1056. doi:10.1098/rsta.2006.1965
26. US Department of Energy, Office of Science 2003 Basic research needs for the hydrogen economy. Report of the basic energy sciences workshop on hydrogen production, storage and use, Washington, DC. Available from http://www.sc.doe.gov/bes/reports/list.html
27. Steinfeld A (2002) Solar hydrogen production via a two-step water-splitting thermochemical cycle based on Zn/ZnO redox reactions. Int J Hydrogen Energy 27:611–619. doi:10.1016/S0360-3199(01)00177-X
28. Vitart X, Duigou AL, Carles P (2006) Hydrogen production using the sulfur-iodine cycle coupled to a VHTR: an overview. Energy Conv Mgmt 47:2740–2747. doi:10.1016/j.enconman.2006.02.010
29. Dincer I, Joshi AS (2013) Solar based hydrogen production systems. Springer, New York. doi:10.1007/978-1-4614-7431-9
30. Kok K (2009) Nuclear engineering handbook. CRC Press, FL. Chapter 5, p 223
31. Ozalp N, Epstein M, Kogan A (2010) Cleaner pathways of hydrogen, carbon nano-materials and metals production via solar thermal processing. J Cleaner Prod 18:900–907. doi:10.1016/j.jclepro.2010.01.020
32. Dutton AG, Bleijs JAM, Dienhart H, Falchetta M, Hug W, Prischich D, Ruddell AJ (2000) Experience in the design, sizing, economics, and implementation of autonomous wind-powered hydrogen production systems. Int J Hydrogen Energy 25:705–722. doi:10.1016/S0360-3199(99)00098-1
33. Ackermann T, Soder L (2000) Wind energy technology and current status: a review. Renew Sust Energy Rev 4:315–374. doi:10.1016/S1364-0321(00)00004-6
34. Saur G, Ramsden T (2011) Wind electrolysis: hydrogen cost optimization. Technical report NREL/TP-5600-50408 Contract no. DE-AC36-08GO28308
35. Melis A (2002) Green Alga hydrogen production: progress, challenges and prospects. Int J Hydrogen Energy 27:1217–1228. doi:10.1016/S0360-3199(02)00110-6
36. Khaselev O, Turner JA (1998) A monolithic photovoltaic-photoelectrochemical device for hydrogen production via water splitting. Science 280:425–427. doi:10.1126/science.280.5362.425
37. Graetzel M (2001) Photoelectrochemical cells. Nature 414:338–344. doi:10.1038/35104607
38. Lewis N (2001) Light work with water. Nature 414:589–590. doi:10.1038/414589a
39. Turner JA (2004) Sustainable hydrogen production. Science 305:972–974. doi:10.1126/science.1103197
40. Bonaquist D (2010) Analysis of CO_2 emissions, reductions, and capture for large-scale hydrogen production plants. Praxair white Paper, Oct 2010. www.praxair.com

Chapter 2
Sodium Hydroxide for Clean Hydrogen Production

Abstract Hydrogen can be generated in several ways utilizing either renewable or non-renewable sources. However, the lack of a clean hydrogen generation methods at a large scale is considered to be one of the obstacles to implement hydrogen economy. The role of sodium hydroxide is increasing as a valuable ingredient to produce hydrogen. However, the vast use of sodium hydroxide is limited due to its (i) corrosive nature and (ii) high-energy-intensive production method. Various current technologies include sodium hydroxide to lower the operating temperature, accelerate hydrogen generation rate as well as sequester carbon dioxide during hydrogen production. Sodium hydroxide finds applications in all the major hydrogen production methods such as steam methane reforming (SMR), coal gasification, biomass gasification, electrolysis, photochemical and thermochemical. Sodium hydroxide, being alkaline, acts as a catalyst, promoter or even a precursor.

Keywords Hydrogen economy · Energy-intensive · Sequester · Steam methane reforming · Coal/biomass gasification · Electrolysis · Photochemical/thermochemical

2.1 Introduction

Hydrogen has a potential to become an environmentally benign energy carrier for the future. However, clean production methods for hydrogen are yet to be identified. Hydrogen's source is either hydrocarbons or water. The primary methods for the generation of hydrogen involve reactions of coal, char, or hydrocarbons with steam at high temperatures. Hydrocarbons are preferred because of their inherent advantages such as their availability, comparable cost, ease of storage and distribution, and relatively high H/C ratio [1]. As stated earlier, water can also produce H_2 (electrolysis process). However, the electrolysis of water is a very energy-intensive process. As burning of fossil fuel is the main energy source for

S. Kumar, *Clean Hydrogen Production Methods*,
SpringerBriefs in Energy, DOI 10.1007/978-3-319-14087-2_2

water electrolysis, splitting of water into its chemical constituents (H_2 and O_2) indirectly leads to vast carbon dioxide emissions. Because of the aforementioned reasons, it seems certain that hydrocarbons are likely to play a significant role in hydrogen production in the near- to medium-term future.

The methods for hydrogen production using hydrocarbon feedstock can be categorized either as follows:

(a) Oxidative (uses oxidants or their combination—O_2, H_2O, CO_2) or
(b) Non-oxidative (splitting of the C–H bond using energy input).

Most of the industrial hydrogen production processes (e.g., steam methane reformation (SMR), partial oxidation, auto-thermal reforming) belong to the oxidative category. The use of oxidants is always accompanied by the vast release of carbon dioxide. A typical oxidant-based hydrogen plant with a capacity of 2.5 million m^3 hydrogen per day vents 1 million m^3 of CO_2 into the atmosphere [1]. Fortunately, the non-oxidative method does not require any oxidant, and therefore, no carbon dioxide is directly emitted via this process. However, the non-oxidative method (e.g., water electrolysis) produces carbon dioxide indirectly through the consumption of fossil-derived electricity.

Of course, the aim is to mitigate carbon emission during hydrogen production. The possible ways are as follows:

(a) use of carbon dioxide absorption unit at a hydrogen plant,
(b) use of nuclear reactors, and/or
(c) thermal dissociation of hydrocarbons into hydrogen and carbon.

However, none of these technologies can completely curb CO_2 emissions and they tend to emit greenhouse gases directly or indirectly. Thus, at present, all these suggested methods have their own limitations. For instance, the integration of carbon dioxide capture unit with hydrogen production process would increase the cost of hydrogen production by about 25–30 % [2]. In the same line, the use of nuclear reactors for hydrogen production gained substantial attention in the last decade because nuclear reactors already produce enough heat for changing water into steam and the electricity for splitting the steam down into hydrogen and oxygen. However, these nuclear reactors are very expensive and cannot be economically feasible to serve world energy demand. As mentioned earlier, another possible way is to thermally dissociate hydrocarbons to produce hydrogen and solid carbon. The thermal dissociation process produces solid carbon, which is far easier to separate compared to gaseous carbon dioxide. It is noteworthy that such method is free of CO_2 separation step, which is highly energy-intensive process. However, at present, the thermal dissociation process itself is fossil-derived and thus cannot eliminate the emission of CO_2.

Currently, there are no other known methods of hydrogen production that do not involve carbon emission, other than those using non-fossil energy. Having this in mind, fossil fuel-based processes that can mitigate carbon dioxide will be worth investigating. Recently, many researchers have proposed the inclusion of sodium hydroxide to the existing hydrogen production technologies. For instance, Reichman et al. [3] suggested a process called Ovonic Renewable Hydrogen (ORH). This method involves sodium hydroxide for the reformation of organic matter to produce hydrogen gas. Moreover, Onwudili and William [4] used sodium hydroxide as a promoter to produce hydrogen gas via hydrothermal gasification of glucose and other biomass samples. Similarly, Kamo et al. [5] pyrolyzed dehydrochlorinated polyvinyl chloride (PVC) and activated carbon with sodium hydroxide and steam to generate hydrogen gas and sodium carbonate with methane, ethane, and carbon dioxide as by-products.

The use of sodium hydroxide for the production of hydrogen has been in application since the nineteenth century. Sodium hydroxide has been proposed as an essential ingredient for most of the present hydrogen producing technologies. Below, sodium hydroxide-based modifications for the methods using fossil fuels, biomass, metal, organic compounds, and water are reviewed. The related concepts and their results are discussed in the following sections.

2.1.1 Overview of Sodium Hydroxide (NaOH)

Previous technology for sodium hydroxide production involved mixing of calcium hydroxide with sodium carbonate. This process was named as "causticizing."

$$Ca(OH)_2(aq.) + Na_2CO_3(s) = CaCO_3 \downarrow + 2NaOH\ (aq.) \qquad (2.1)$$

Currently, sodium hydroxide is produced by the electrolysis of brine (NaCl):

$$2NaCl + 2H_2O = 2NaOH + Cl_2 \uparrow + H_2 \uparrow \qquad (2.2)$$

Reaction (2.2) is an energy-intensive process and thus has significant carbon footprint. Besides sodium hydroxide, reaction (2.2) also produces toxic chlorine and hydrogen as by-products. It is necessary to modify reaction (2.2) in such a way that can significantly reduce the emission of CO_2 and toxic chlorine to the atmosphere.

Figure 2.1 illustrates the membrane cell used for the electrolysis of brine [6]. The three commercially available production methods for sodium hydroxide are compared here (Table 2.1). The strength of soda solutions and amount of required steam varies for different production methods. As a diaphragm cell produces the least concentrated soda solutions, evaporation is required to raise the concentration up to 50 wt% solution as in mercury cell process.

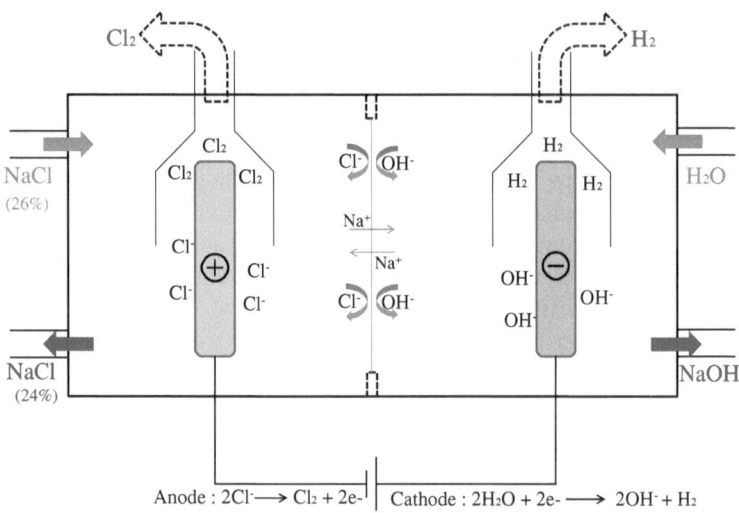

Fig. 2.1 Membrane cell process schematic for the production of sodium hydroxide previously published in [6], with permission from Formatex Research Center © 2013

Table 2.1 Comparison of the commercially available production methods for NaOH [81]

Factors	Diaphragm	Mercury	Membrane
Use of mercury	No	Yes	No
Chlorine as a by-product	Yes	No	Yes
Operating current density (kA/m^2)	0.9–2.6	8–13	3–5
Cell voltage (V)	2.9–3.5	3.9–4.2	3.0–3.6
NaOH strength (wt%)	12	50	33–35
Energy consumption (kWh/Mt Cl$_2$) at a current density (kA/m^2)	2,720 (1.7)	3,360 (10)	2,650 (5)
Steam consumption (kWh/MT Cl$_2$) for concentration to 50 % NaOH	610	0	180
% NaOH produced in USA	62	10	24

2.2 Hydrogen Production

2.2.1 Fossil Fuels

Currently, SMR is the most common and the least expensive industrial technology to produce hydrogen [7]. Methane reacts at a high temperature (700–1,100) °C with steam to form syngas (CO + H$_2$).

$$CH_4(g) + H_2O(g) = CO(g) + 3H_2(g) \Delta H_{1227\,°C} = 397 \text{ kJ/mol} \qquad (2.3)$$

Syngas can further react with steam to form additional hydrogen at a lower temperature.

$$CO(g) + H_2O(g) = CO_2(g) + H_2(g) \Delta H_{327\,°C} = -242 \text{ kJ/mol} \qquad (2.4)$$

Thus, the combined reaction is

$$CH_4(g) + 2H_2O(g) = CO_2(g) + 4H_2(g) \Delta H_{927\,°C} = 431 \text{ kJ/mol} \qquad (2.5)$$

The enthalpy change (ΔH) is provided for temperatures at which the reaction is producing the maximum hydrogen and calculated using FACTSAGE^{TM} software. It should be noted that the equilibrium values of the gases are different from what is given by stoichiometric proportions of species on the product side of the equations.

Figure 2.2 depicts the simplified block diagram of SMR technique equipped with carbon dioxide absorption unit and a methanation reactor. The main operating units include natural gas feedstock desulphurization, catalytic reforming, water–gas shift reactor, and CO_2 gas separation and hydrogen purification [8].

In the desulfurization unit, sulfur-based organic compounds (such as thiols) are first converted into H_2S by catalytic hydrogenation reaction (Co–Mo catalysts, 290–370 °C) [9]. Further, H_2S reacts with ZnO to form ZnS. (H_2S + ZnO $\rightarrow$ ZnS + H_2O, 340–390 °C). Natural gas feedstock must be pretreated before mixing with steam (2.6 MPa). And then, the mixture should be heated to 500 °C prior to sending the SMR unit. SMR is favored by low pressure and performed in the reactor at usually 2.0–2.6 MPa. The gaseous mixture (H_2, CO and steam) leaves the reformer at 800–900 °C. It is cooled rapidly to 350 °C and fed to the water–gas shift

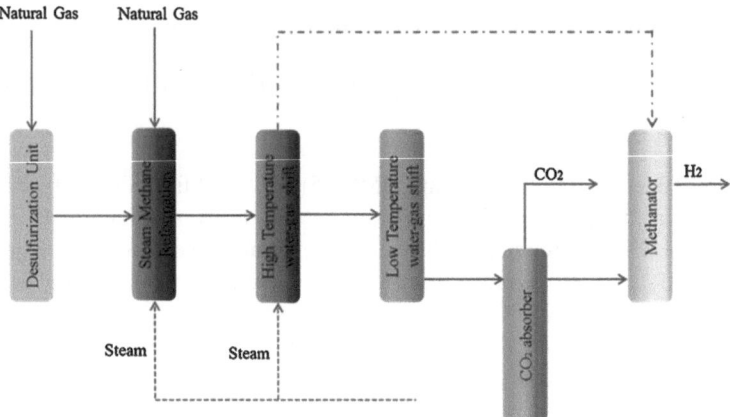

Fig. 2.2 A schematic of hydrogen production by SMR—with solvent removal of CO_2 and a methanation unit [8]

reactor, where reaction (2.4) is operated. Reaction (2.4) produces hydrogen and carbon dioxide. Carbon dioxide is captured using amine-based solvent—mono-ethanolamine (MEA). And the remaining residuals, CO_2 and CO, are fed to methanation reactor, where the mixture is converted to CH_4 in the presence of hydrogen (320 °C, Ni or Ru on oxide support as catalyst) [8].

Interestingly, a simple calculation shows that about 10.5 g of CO_2 is emitted per gram of H_2 production via SMR technique. Such an undesired vast emission of CO_2 endangers the prolonged use of conventional SMR technique to produce hydrogen. However, as mentioned earlier, the integration of an amine-based CO_2 capture unit to hydrogen production process would significantly increase the cost of hydrogen per ton. Therefore, at present, most of the SMR sites are not equipped with CO_2 absorber unit.

In this regard, several new methods are proposed that could solve the existing problems [10–15]. Berthelot described the reaction between NaOH and CO which yields sodium formate (HCOONa). When heated above 250 °C, HCOONa transforms into oxalate with the release of hydrogen:

$$NaOH\ (s) + CO(g) = HCOONa\ (s) \tag{2.6}$$

$$2HCOONa(s) = Na_2C_2O_4(s) + H_2(g) \tag{2.7}$$

In 1918, Boswell and Dickson demonstrated that when carbon monoxide is heated with excess of sodium hydroxide at temperatures at which formate is transformed into oxalate, oxidation almost quantitatively convert to carbon dioxide occurs with the evolution of an equivalent amount of hydrogen [16]:

$$2\ NaOH(s) + CO(g) = Na_2CO_3(s) + H_2(g) \tag{2.8}$$

Similarly, Saxena proposed the inclusion of sodium hydroxide as an additional reactant to the conventional SMR system. The addition of sodium hydroxide serves the dual purpose of carbon sequestration and H_2 production [17].

$$2NaOH(s) + CH_4(g) + H_2O(g) = Na_2CO_3(s) + 4H_2(g) \quad \Delta H_{427\,°C} = 244\ kJ/mol \tag{2.9}$$

Figure 2.3 compares the standard SMR (5) and modified SMR (9). It can be observed from the phase equilibrium diagram that unlike modified SMR method, conventional SMR technique produces a more complex composition of gas (CO, CO_2, H_2O, H_2) and also requires comparatively more energy (431 kJ/mol at 927 °C versus 244 kJ/mol at 427 °C).

Coal gasification is another well-established technology to produce hydrogen. However, it is also an energy-intensive process. Here, oxygen or steam is passed over coal to produce a gaseous mixture of H_2, CO, and CO_2 from which H_2 is separated (Fig. 2.4).

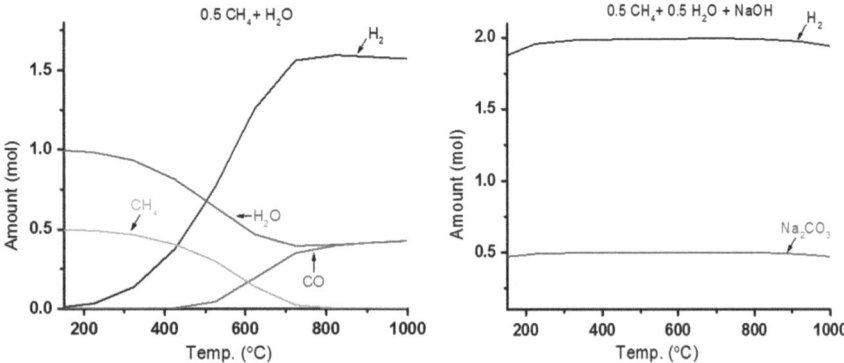

Fig. 2.3 Calculated equilibrium in the system **a** conventional SMR and **b** modified SMR reactions

Fig. 2.4 Production of hydrogen from coal gasification [1]

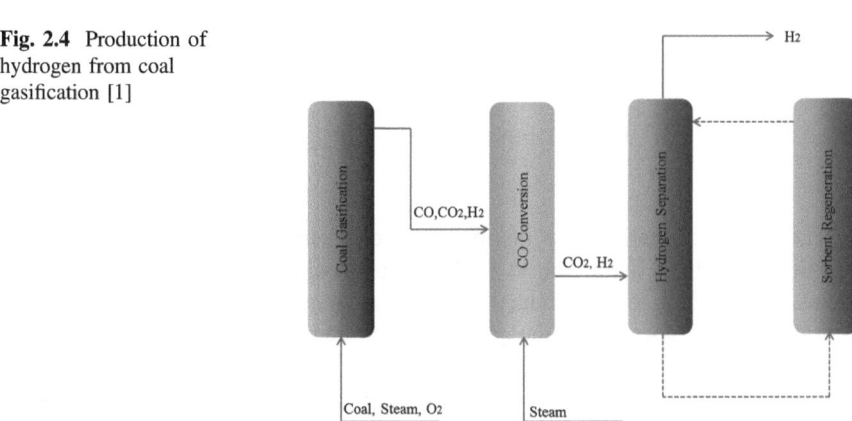

$$2C(s) + 3H_2O(g) \rightarrow CO(g) + CO_2(g) + 3H_2(g) \quad \Delta H_{327\,°C} = 95.73 \text{ kJ/mol} \tag{2.10}$$

Coal-made hydrogen has applications in the production of ammonia, methanol, methane, and Fischer–Tropsch products. However, coal gasification suffers critical limitations [1], and it is given as follows:

(a) not as cost-effective as producing hydrogen from oil or natural gas;
(b) an endothermic reaction; and
(c) with vast emission of CO_2.

Figure 2.5 illustrates the thermodynamic calculation on [NaOH(s) + C(s) + H$_2$O (g)] using Factsage^{TM} software. The inclusion of sodium hydroxide to the coal–steam system can significantly reduce the energy input (95.73 kJ/mol reduced to 64.58 kJ/mol at 327 °C) [18]. The system not only captures CO_2 in the form of soda ash but also produces hydrogen. Moreover, the system does not produce complex mixture of gases.

Fig. 2.5 Calculated
equilibrium in the system
$2NaOH(s) + C(s) + H_2O(g)$

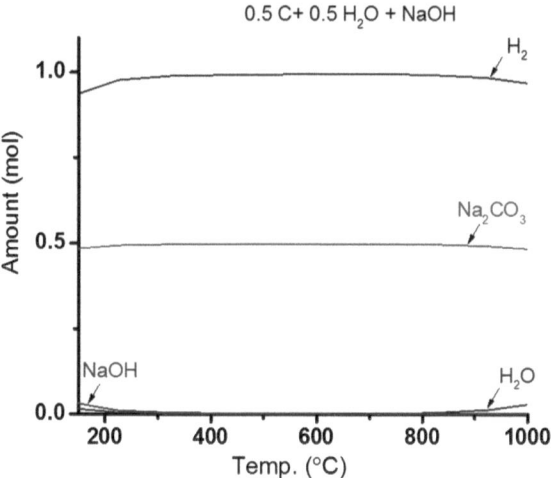

$$2NaOH(s) + C(g) + H_2O(g) = Na_2CO_3(s) + 2H_2(g) \quad \Delta H_{327^\circ C} = 64.58 \text{ kJ/mol}$$
$$(2.11)$$

Table 2.2 summarizes the thermodynamic calculation and the effect for the addition of sodium hydroxide to methane and coal. It can be observed that the inclusion of alkali reduces both the operating temperature and carbon dioxide emission. As a consequence, the amount of coal required to run these processes is also reduced. Table 2.2 summarizes the inclusion of sodium hydroxide to CH_4 and C in the presence of steam. Sodium hydroxide captures CO_2 and forms sodium carbonate (Na_2CO_3), which has huge application in different chemical sectors such as glass manufacturing, electrolyte, textiles, and domestic use.

Sodium hydroxide is already in use for hydrogen production at industrial scale. For instance, the black liquor gasification process utilizes alkali hydroxide to serve the dual purpose of hydrogen production and carbon sequestration. In a typical pulping process for paper production, approximately one-half of the raw materials are converted to pulp and other half are dissolved in the black liquor. The black liquor solution consists of well-dispersed carbonaceous material, steam, and alkali

Table 2.2 Thermodynamic properties for different hydrogen production methods after inclusion of NaOH [82]

	$CH_4 + H_2O$	$NaOH + CH_4 + H_2O$	$C + H_2O$	$NaOH + C + H_2O$
Temperature (°C)	700–1,100	600–800	800–1,200	500–700
Enthalpy (ΔH, kJ/mol)	431 (927 °C)	244 (427 °C)	95.73 (327 °C)	64.58 (327 °C)
Mixture of product gases	CO, CO_2, H_2	H_2	CO, CO_2, H_2	H_2
Coal/H_2 (g/g)	1.64	0.93	3.73	3.49
CO_2/H_2 (g/g)	10.5	3.41	13.67	1.80

metal which are burned to provide part of energy for the plant. Due to the presence of carbonaceous material and water in the liquor, the following carbon–water reaction dominates:

$$C(s) + H_2O(g) = CO(g) + H_2(g) \tag{2.12}$$

$$CO(g) + H_2O(g) = CO_2(g) + H_2(g) \tag{2.13}$$

However, due to the thermodynamic limitations, reaction (2.13) never proceeds toward completion; therefore, hydrogen concentration does not exceed a certain limit. Interestingly, in the presence of NaOH, CO_2 capture medium, the equilibrium can be shifted to drive reaction (2.13) toward completion and therefore maximize hydrogen concentration. Consequently, the concentration of CO and CO_2 is reduced significantly in the product gases.

It is noteworthy that the modifications based on the use of sodium hydroxide cannot be considered as a solution on the global scale. Sodium hydroxide itself is produced using electrolysis of brine which is a highly energy-intensive process.

2.2.2 Biomass

Biomass is a renewable energy source and is regarded as a carbon-neutral fuel. It consumes the same amount of carbon while growing as much it releases when burnt as a fuel.

Biomass gasification means incomplete combustion of biomass resulting in the production of combustible gases consisting of CO, H_2, and traces of CH_4.

$$\begin{aligned} \text{Biomass} + \text{heat} + \text{steam} &\rightarrow H_2 + CO + CO_2 + CH_4 \\ &+ \text{Light/Heavy hydrocarbons} + \text{Char} \end{aligned} \tag{2.14}$$

The major challenges that the gasification process mainly faces are as follows: (1) undesirable tar formation and (2) huge carbon emission. The tar may cause the formation of tar aerosol and a more complex polymer structure, which are not favorable for hydrogen production through steam reformation. The suggested solutions to minimize tar formation are as follows:

(a) proper designing of gasifier;
(b) proper control and operation; and
(c) use of additives or catalysts.

The addition of sodium hydroxide to biomass gasification can solve some of the existing problems [19]. Firstly, inclusion of sodium hydroxide can reduce the carbon emission. Cellulose [$C_6H_{10}O_5$], D-glucose [$C_6H_{12}O_6$], and sucrose [$C_{12}H_{22}O_{11}$] react with water vapor in the presence of sodium hydroxide to form sodium carbonate and hydrogen. The mechanism of the alkali-promoted reaction suggests that the dehydrogenation of cellulose in presence of Na^+ and OH^- ions yields hydrogen. The concentration of Na^+ and OH^- ions strongly influences the dehydrogenation of cellulose [20, 21]. However, the product also consists of hydrocarbons such as CH_4 and lowers the percentage of hydrogen yield. But nickel catalysts supported on alumina can reduce the formation of CH_4 and maximize the hydrogen yield [4, 20–22].

Secondly, sodium hydroxide can also reduce the pyrolysis temperature of biomass species [23]. Sodium ions, being small, can penetrate into the biomass texture and break the hydrogen bridges. Consequently, devolatilization occurs rapidly. Su et al. used a new catalyst derived from sodium aluminum oxide ($Al_2O_3.Na_2O$), $Al_2O_3.Na_2O.xH_2O/NaOH/Al(OH)_3$, to increase the hydrogen yield for steam gasification of cellulose. The gasification temperature was kept below 500 °C to prevent any tar formation [24, 25].

Sodium hydroxide acts as a promoter of hydrogen gas during the hydrothermal gasification of glucose and other biomass samples. However, the cost of alkali metal and its proper recycling are major concerns for the use of sodium hydroxide in the biomass gasification.

2.2.3 Metals

Metals can react with sodium hydroxide in the presence or absence of water to produce hydrogen. Transition metal reacts with sodium hydroxide to form metal oxides and hydrogen [26]. Ferrosilicon too generates hydrogen on reacting with sodium hydroxide [27].

Here, the reaction of aluminum (the most abundant metal in Earth's crust) is considered with sodium hydroxide in the presence of water vapor. As can be expected, hydrogen gas is generated from the chemical reaction between Al and water (3.7 wt% H_2, theoretical yield) [28]. Al/H_2O system is indeed a safe method to generate hydrogen. But the system has kinetic limitations as the metal surface passivation in neutral water occurs more easily and the metal activity with water is extremely low. Thus, aluminum activity in water needs to be improved. To solve the problem of surface passivation of Al, various solutions have been suggested so far. The solutions either include the addition of hydroxides [29, 30], metal oxides [31, 32], selected salts [33, 34], or alloying Al with low melting point metal [35–38]. The alkali-promoted Al/H_2O system is favored over other metal systems because of the high hydrogen generation rate.

The reaction between Al and H_2O with sodium hydroxide solution produces hydrogen, which can be expressed as follows:

$$2Al + 6H_2O + NaOH \rightarrow 2NaAl(OH)_4 \downarrow + 3H_2 \uparrow \qquad (2.15)$$

$$NaAl(OH)_4 \rightarrow NaOH + Al(OH)_3 \downarrow \qquad (2.16)$$

Sodium hydroxide consumed for the hydrogen generation in exothermic reaction (2.15) will be regenerated through the decomposition of $NaAl(OH)_4$ via reaction (2.16). Reaction (2.16) also produces a crystalline precipitate of aluminum hydroxide. The combination of above two reactions completes the cycle and demonstrates that only water will be consumed in the whole process if the process is properly monitored. Some of the previous work conducted in this sphere reported kinetics of the reaction between Al and H_2O with sodium hydroxide solution and calculated the activation energy in the range of 42.5–68.4 kJ/mol [39, 40].

Many researchers examined the effects of other crucial parameters that control the hydrogen generation properties for alkali-assisted Al/H_2O system. The parameters include temperature, alkali concentration, morphology, initial amount of Al, and concentration of aluminate ions [41, 42]. Moreover, Soler et al. compared the hydrogen generation performance for three different hydroxides: NaOH, KOH, and $Ca(OH)_2$. They observed that sodium hydroxide solution consumes Al faster compared with other two hydroxides [42]. Similarly, S.S. Martinez et al. treated Al-can wastes with NaOH solution at room temperature to generate pure hydrogen. The by-product $(NaAl(OH)_4)$ was used to prepare a gel of $Al(OH)_3$ to treat drinking water contaminated with arsenic [43].

2.2.4 Water-Splitting Thermochemical Cycle

Water is a basic source of hydrogen. However, direct splitting of water in hydrogen and oxygen requires huge amount of energy. Therefore, researchers are exploring innovative methods to resolve such issue. One interesting concept is the utilization of renewable sources (such as solar energy) to split water in the presence of metal oxides [44–46]. Figure 2.6 illustrates this concept, which commonly known as water-splitting thermochemical cycles.

The figure conveys a three-step water-splitting process:

1. reduction of oxides (energy-intensive process, 800–1,000 °C)
2. reaction of reduced oxide with sodium hydroxide (hydrogen generation step), and
3. hydrolysis reaction (sodium hydroxide recovery step).

$$(I)\ MO(ox) = MO(red) + 0.5O_2 \qquad (2.17)$$

$$(II)\ MO(red) + 2\ NaOH = Na_2O \cdot MO(ox) + H_2 \qquad (2.18)$$

$$(III)\ Na_2O \cdot MO(ox) + H_2O = MO(ox) + 2NaOH \qquad (2.19)$$

Any thermodynamically favorable oxide can be selected to generate hydrogen. Thus, so far, a large number of oxides have been considered. The water-splitting

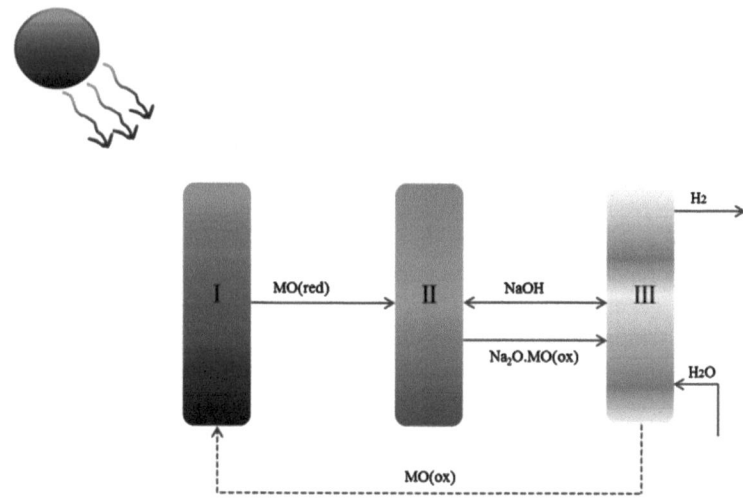

Fig. 2.6 Schematic for water-splitting thermochemical cycle *MO* metal oxide

thermochemical cycle reactions can be mainly classified as (1) two-step water-splitting process [47–51] (2) iodine–sulfur process [52–54] and (3) calcium–bromine process [55–57].

Table 2.3 summarizes the findings for various alkali metals used for water-splitting thermochemical cycles. Sodium hydroxide is able to generate hydrogen at a reduced temperature. Recently, Miyoka et al. [58] considered sodium redox reaction and conducted several experiments in a non-equilibrium condition but could not achieve a 100 % conversion. It was attributed to the slow kinetics of both the hydrogen generation reaction and sodium recovery. Moreover, sodium hydroxide facilitates oxidation in the water-splitting step. But the volatility of sodium hydroxide at temperatures higher than 800 °C and incomplete Na$^+$ extraction by water to recover sodium hydroxide limit its application. Several research groups concluded that even though sodium- or sodium hydroxide-assisted reaction has major advantages, their recovery could be a big challenge. In the same line, Weimer et al. recommend membrane separation to recover sodium hydroxide [59]. In recent times, few researchers also investigated the use of Na$_2$CO$_3$ as an alternative to NaOH [60–62].

Besides sodium hydroxide recovery, there are other limitations too. For instance, the reduction of oxides requires very high temperature. To attain such high temperature, a large-scale solar heat plant will be required. At present, the construction of a large thermochemical hydrogen plants is limited by the location, cost, and safety issues. Therefore, techniques to lower the operation temperature of water-splitting process should be investigated. A low temperature water-splitting process will allow the utilization of small-scale solar heat systems or even exhaust heat from industries. As sodium hydroxide can significantly reduce the operation temperature of the water-splitting process, it could be considered for such use.

Table 2.3 Alkali metal-assisted water-splitting thermochemical cycle

System	Reactions	Conditions	Hydrogen yield	Remarks	Ref.
MnO	$Mn_2O_3 = 2MnO + 0.5O_2$ (<1,600 °C) $2MnO + 2NaOH = H_2 + 2\alpha\text{-}NaMnO_2$ (~700 °C) $2\alpha\text{-}NaMnO_2 + H_2O = Mn_2O_3 + 2NaOH$ (<100 °C) $H_2O = H_2 + 0.5O_2$	Hydrogen generation at 750 °C	100 % conversion under vacuum (0.5 h) and under N_2 purge (3 h)	NaOH recovery improved from 10 to 35 % in (MnO + Fe) mixture. Difficult Mn_2O_3–NaOH separation	[83]
Ce$_2$Ti$_2$O$_7$, Ce$_2$Si$_2$O$_7$, CeFeO$_3$, CeVO$_4$, CeNbO$_4$	$MO(ox) = MO(red) + 0.5\ O_2$ $MO(red) + 2\ M'OH = M'_2O.MO(ox) + H_2$ $M'_2O.MO(ox) + H_2O = MO\ (ox) + 2\ M'OH$	Mixed oxide synthesis around 1,500 °C Hydrogen generation T range 500–800 °C	(1.5–1.94) mmol/g oxide	At 530 °C, Ce$_2$Si$_2$O$_7$ (highest reaction efficiency) Hydrogen generation infeasible up to 1,000 °C	[84]
Zn–Mn–O	$Zn_{0.66}Mn_2O_{3.66} = 2Zn_{0.33}MnO_{1.33} + 0.5O_2$ (~1,600 °C) $2Zn_{0.33}MnO_{1.33} + 2NaOH = H_2 + Na_2Zn_{0.66}Mn_2O_{4.66}$ (>650 °C) $Na_2Zn_{0.66}Mn_2O_{4.66} + H_2O = Zn_{0.66}Mn_2O_{3.66} + 2NaOH$ (<100 °C)	Hydrogen generation above 650 °C	80–90 % conversion rate under low pressure and residence time of 0.5 h	NaOH may be recovered using membrane process	[59]
NaOH	$2NaOH + Na = Na_2O + H_2$ ($T_{eq} = 32$ °C) $2Na_2O = Na_2O_2 + 2Na$ ($T_{eq} = 1,870$ °C) $Na_2O_2 + H_2O = 2NaOH + 0.5O_2$ (100 °C)	Non-equilibrium technique for hydrogen production, below 400 °C	>80 % at 350 °C	Yield of H$_2$ generation and Na separation <100 %, kinetic limitation, suitable catalysts need to be investigated	[58]

2.2.5 Organic Compounds

2.2.5.1 Formic Acid (HCOOH)

Formic acid and its solution are industrial hazards. Any use of such chemical waste will be of a great advantage for environment. Formic acid can produce hydrogen using two methods: (1) thermo catalytic decomposition and (2) electrolysis in the presence of sodium hydroxide.

Formic acid thermally decomposes to produce H_2 and CO_2 [HCOOH (l) $\rightarrow$ H_2 (g) + CO_2 (g), $\Delta G° = -32.9$ kJ/mol, $\Delta H° = 31.2$ kJ/mol)], the reversible reaction of CO_2 hydrogenation [63–71]. Electrolysis of formic acid solutions in the presence of sodium hydroxide requires theoretically much lower energy than water [72]. The electrochemical reaction for the electrolysis of formic acid solutions is as follows [73]:

$$\text{Anode:} \qquad HCOOH + OH^- \rightarrow CO_2 + H_2O + 2e^-$$

$$\text{Cathode:} \qquad 2H_2O + 2e^- \rightarrow H_2 + 2OH^- \qquad (2.20)$$

$$\text{Overall reaction:} \qquad HCOOH \rightarrow H_2 + CO_2$$

Figure 2.7 demonstrates the scheme of electricity generation via the combined use of alkaline hydroxide (sodium hydroxide) for the electrolysis of formic acid (HCOOH) and fuel cell. The separation of H_2 and CO_2 is desired prior to injection in the fuel cell.

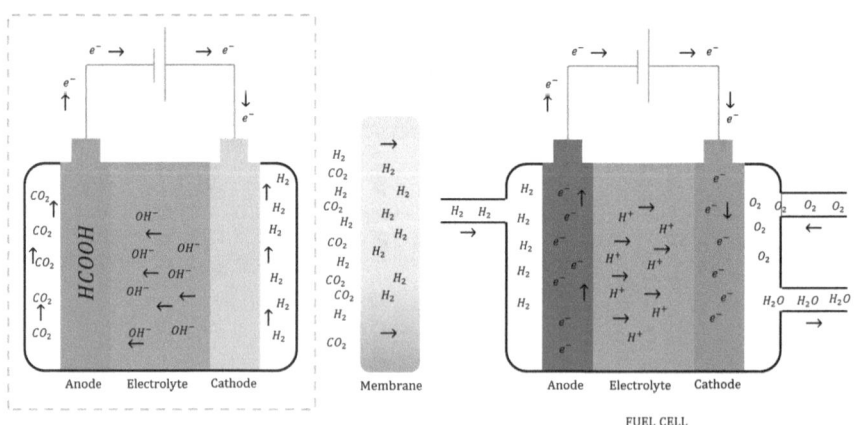

Fig. 2.7 Electricity generation using alkaline hydroxide (NaOH) for the electrolysis of HCOOH

2.2.5.2 Formaldehyde (HCHO)

An aqueous solution of formaldehyde reacts with sodium hydroxide to produce small amounts of hydrogen [74]. The generation of hydrogen competes with the disproportionation of formaldehyde to form corresponding alcohol and acid [75, 76]. Further, Ashby et al. [77] proposed a mechanistic explanation of this process. The mechanism indicates that one hydrogen atom originates from the water and the other from the organic moiety. According to an experimental study, when a dilute solution of formaldehyde (4×10^{-4} m) reacts with concentrated sodium hydroxide (19 m) at room temperature, hydrogen is produced in a significant amount [77]. However, when concentrated solution of formaldehyde interacts with dilute sodium hydroxide solution, only a trace amount of hydrogen is produced.

Moreover, when a solution of hydrogen peroxide mixes with formaldehyde and sodium hydroxide, hydrogen is generated again [78]. Hydrogen peroxide oxidizes formaldehyde to formic acid and sodium hydroxide further neutralizes the acid.

$$H_2O_2 + 2HCHO + 2NaOH = 2HCOONa + H_2 + 2H_2O \qquad (2.21)$$

However, no trace of hydrogen is observed in the absence of sodium hydroxide [79, 80]. The reaction (2.21) is limited by slow kinetics and requires a large excess of alkali hydroxide. When hydrogen peroxide is replaced by cuprous oxide, hydrogen is generated in a quantitative amount.

2.3 Conclusion

Sodium hydroxide locks CO_2 in the form of valuable chemical compound, Na_2CO_3. The use of sodium hydroxide for the production of hydrogen results in high hydrogen generation rates, lower operation temperatures, and overall reduction in carbon emission. Apparently, sodium hydroxide has significant role to play for the methods using either renewable or non-renewable energy sources. But the energy-intensive production method of sodium hydroxide (electrolysis of brine) limits its application at a large scale. Thus, it will be of a great interest to invent or modify a method that can produce sodium hydroxide using renewable resources such as solar energy, water, and wind. Moreover, new hydrogen generation concepts that can replace sodium hydroxide with industrial hazards or waste should also be explored. In the following chapter, a detail for the role of sodium hydroxide to the industrial hydrogen production technologies such as SMR and coal gasification is provided. The chapter also presents the effect of different catalysts over the kinetics of these modified reactions.

References

1. Gupta RB (2009) Hydrogen fuel: production, transport, and storage, Chapter 2. CRC Press, Boca Raton
2. Audus H, Kaarstad O, Kowal M (1996) Decarbonization of fossil fuels: hydrogen as an energy vector. In: Proceedings of 11th world hydrogen energy conference. Stuttgart, Germany
3. Reichman B, Mays W, Strebe J, Fetcenko M (2010) Ovonic renewable hydrogen (ORH)—low temperature hydrogen from renewable fuels. Int J Hydrogen Energy 35:4918–4924. doi:10.1016/j.ijhydene.2009.08.097
4. Onwudili JA, Williams PT (2009) Role of sodium hydroxide in the production of hydrogen gas from the hydrothermal gasification of biomass. Int J Hydrogen Energy 34:5645–5656. doi:10.1016/j.ijhydene.2009.05.082
5. Kamo T, Takaoka K, Otomo J, Takahashi H (2006) Effect of steam and sodium hydroxide for the production of hydrogen on gasification of dehydrochlorinated poly(vinyl) chloride. Fuel 85:1052–1059. doi:10.1016/j.fuel.2005.10.002
6. Kumar S, Saxena SK (2013) Role of sodium hydroxide for hydrogen gas production and storage. In: Mendez-Vilas A (ed) Materials and processes for energy: communicating current research and technological developments. Formatex Research Center, Spain
7. Probstein RF, Hicks RE (2000) Synthetic fuels, Chap 2. Dover, New York
8. Muradov N (2009) Production of hydrogen from hydrocarbons. In: Gupta R (ed) Hydrogen fuel, production, transport and storage. Boca Raton, FL
9. Armor J (1999) The multiple roles for catalysis in the production of H_2. Appl Catal A General 176:159–176. doi:10.1016/S0926-860X(98)00244-0
10. Gupta H, Mahesh I, Bartev S, Fan LS (2004) Enhanced hydrogen production integrated with CO_2 separation in a single-stage reactor; DOE contract no: DE FC26-03NT41853. Columbus
11. Ziock H-J, Lackner KS, Harrison DP (2001) Zero emission coal power, a new concept. In: Proceedings of 1st national conference on carbon sequestration. Washington
12. Stowinski G (2006) Some technical issues of zero-emission coal technology. Int J Hydrogen Energy 31:1091–1102. doi:10.1016/j.ijhydene.2005.08.012
13. Cormos CC, Starr F, Tzimas E, Peteves S (2008) Innovative concept for hydrogen production processes based on coal gasification with CO_2 capture. Int J Hydrogen Energy 33:1286–1294. doi:10.1016/j.ijhydene.2007.12.048
14. Chiesa P, Consonni S, Kreutz T, Williams R (2005) Co-production of hydrogen, electricity and CO_2 from coal with commercially ready technology. Part A: performance and emissions. Int J Hydrogen Energy 30:747–767. doi:10.1016/j.ijhydene.2004.08.002
15. Wang Z, Zhou J, Wang Q, Fan J, Cen K (2006) Thermodynamic equilibrium analysis of hydrogen production by coal based on Coal/CaO/H_2O gasification system. Int J Hydrogen Energy 31:945–952. doi:10.1016/j.ijhydene.2005.07.010
16. Boswell MC, Dickson JV (1918) The fusion of sodium hydroxide with some inorganic salts. J Am Chem Soc 40:1779–1786. doi:10.1021/ja02245a003
17. Saxena S, Kumar S, Drozd V (2011) A modified steam-methane-reformation reaction for hydrogen production. Int J Hydrogen Energy 36:4366–4369. doi:10.1016/j.ijhydene.2010.12.133
18. Saxena S, Drozd V, Durygin A (2008) A fossil-fuel based recipe for clean energy. Int J Hyd Energy 33:3625–3631. doi:10.1016/j.ijhydene.2008.04.050
19. Ishida M, Takenaka S, Yamanaka I, Otsuka K (2006) Production of CO_x-free hydrogen from biomass and NaOH mixture: effect of catalysts. Energy Fuels 20:748–753. doi:10.1021/ef050282u
20. Minowa T, Fang Z, Ogi T, Varhegyi G (1998) Decomposition of cellulose and glucose in hot-compressed water under catalyst-free conditions. J Chem Eng Jpn 31:131–134. doi:10.1252/jcej.31.131

21. Muangrat R, Onwudili JA, Williams PT (2010) Alkali-promoted hydrothermal gasification of biomass food processing waste: a parametric study. Int J Hydrogen Energy 35:7405–7415. doi:10.1016/j.ijhydene.2010.04.179

22. Minowa T, Fang Z (1998) Hydrogen production from cellulose in hot compressed water using reduced nickel catalyst: product distribution at different reaction temperatures. J Chem Eng Jpn 31:488–491. doi:10.1016/S0920-5861(98)00277-6

23. Wang J, Zhang M, Chen M, Min F, Zhang S, Ren Z, Yan Y (2006) Catalytic effects of six inorganic compounds on pyrolysis of three kinds of biomass. Thermochim Acta 444:110–114. doi:10.1016/j.tca.2006.02.007

24. Su S, Li W, Bai Z, Xiang H, Bai J (2010) Production of hydrogen by steam gasification from lignin with $Al_2O_3 \cdot Na_2O \cdot xH_2O/NaOH/Al(OH)_3$ catalyst. J Fuel Chem Technol 238:270–274. doi:10.1016/S1872-5813(10)60032-1

25. Su S, Li W, Bai Z, Xiang H (2008) A preliminary study of a novel catalyst $Al_2O_3 \cdot Na_2O \cdot xH_2O/NaOH/Al(OH)_3$ for production of hydrogen and hydrogen-rich gas by steam gasification from cellulose. Int J Hydrogen Energy 33:6947–6952. doi:10.1016/j.ijhydene.2008.09.003

26. Williams DD, Grand JA, Miller RR (1956) The reactions of molten sodium hydroxide with various metals. J Am Chem Soc 78:5150–5155. doi:10.1021/ja01601a004

27. Annual Report National Advisory Committee for aeronautics (1934) Washington

28. Wang HZ, Leung DYC, Leung MKH, Ni M (2009) A review on hydrogen production using aluminum and aluminum alloys. Renew Sustain Energy Rev 13:845–853. doi:10.1016/j.rser.2008.02.009

29. Belitskus D (1970) Reaction of aluminum with sodium hydroxide solution as a source of hydrogen. J Electrochem Soc 117:1097–1099. doi:10.1149/1.2407730

30. Jung CR, Kundu A, Ku B, Gil JH, Lee HR, Jang JH (2008) Hydrogen from aluminum in a flow reactor for fuel cell applications. J Power Sources 175:490–494. doi:10.1016/j.jpowsour.2007.09.064

31. Deng ZY, Tang YB, Zhu LL, Sakka Y, Ye J (2010) Effect of different modification agents on hydrogen-generation by the reaction of Al with water. Int J Hydrogen Energy 35:9561–9568. doi:10.1016/j.ijhydene.2010.07.027

32. Dupiano P, Stamatis D, Dreizin EL (2011) Hydrogen production by reacting water with mechanically milled composite aluminum metal oxide powders. Int J Hydrogen Energy 36:4781–4791. doi:10.1016/j.ijhydene.2011.01.062

33. Skrovan J, Alfantazi A, Troczynski T (2009) Enhancing aluminum corrosion in water. J Appl Electrochem 39:1695–1702. doi:10.1007/s10800-009-9862-x

34. Soler L, Macana's J, Mun˜oz M, Casado J (2005) Hydrogen generation from aluminum in a non-consumable potassium hydroxide solution. In: Proceedings of international hydrogen energy congress and exhibition IHEC. Istanbul, Turkey

35. Wang W, Chen DM, Yang K (2010) Investigation on microstructure and hydrogen generation performance of Al-rich alloys. Int J Hydrogen Energy 35:12011–12019. doi:10.1016/j.ijhydene.2010.08.089

36. Ziebarth JT, Woodall JM, Kramer RA, Choi G (2011) Liquid phase enabled reaction of Al–Ga and Al–Ga–In–Sn alloys with water. Int J Hydrogen Energy 36:5271–5279. doi:10.1016/j.ijhydene.2011.01.127

37. Fan MQ, Xu F, Sun LX (2007) Hydrogen generation by hydrolysis reaction of ball-milled Al–Bi alloys. Energy Fuels 21:2294–2298. doi:10.1021/ef0700127

38. Ilyukhina AV, Kravchenko OV, Bulychev BM, Shkolnikov EI (2010) Mechanochemical activation of aluminum with galliams for hydrogen evolution from water. Int J Hydrogen Energy 35:1905–1910. doi:10.1016/j.ijhydene.2009.12.118

39. Aleksandrov YA, Tsyganova EI, Pisarev AL (2003) Reaction of aluminum with dilute aqueous NaOH solutions. Russ J General Chem 73:689–694. doi:10.1023/A:1026114331597

40. Zhuk AZ, Sheindlin AE, Kleymenov BV (2006) Use of low-cost aluminum in electric energy production. J Power Sour 157:921–926. doi:10.1016/j.jpowsour.2005.11.097

41. Stockburger D, Stannard JH, Rao BML, Kobasz W, Tuck CD (1992) On-line hydrogen generation from aluminum in an alkaline solution. In: Proc Symp Hydrogen Storage Mater, Batteries Electrochem 92:431–444
42. Soler L, Macana's J, Mun~oz M, Casado J (2007) Aluminum and aluminum alloys as sources of hydrogen for fuel cell applications. J Power Sour 169:144–149. doi:10.1016/j.jpowsour.2007.01.080
43. Martı'nez SS, Benı'tesa WL, Gallegosa A, Sebastia'n PJ (2005) Recycling of aluminum to produce green energy. Solar Energy Mater Solar Cells 88:237–243. doi:10.1016/j.jsolmat.2004.09.022
44. Yalcin S (1989) A review of nuclear hydrogen production. Int J Hydrogen Energy 14:551–561. doi:10.1016/0360-3199(89)90113-4
45. Abanades S, Charvin P, Flamant G, Neveu P (2006) Screening of water-splitting thermochemical cycles potentially attractive for hydrogen production by concentrated solar energy. Energy 31:2805–2822. doi:10.1016/j.energy.2005.11.002
46. Holladay JD, Hu J, King DL, Wang Y (2009) An overview of hydrogen production technologies. Catal Today 139:244–260. doi:10.1016/j.cattod.2008.08.039
47. Nakamura T (1977) Hydrogen production from water utilizing solar heat at high temperatures. Sol Energy 19:467–475. doi:10.1016/0038-092X(77)90102-5
48. Sibieude F, Ducarroir M, Tofighi A, Ambriz J (1982) High temperature experiments with a solar furnace: the decomposition of Fe_3O_4, Mn_3O_4 CdO. Int J Hydrogen Energy 7:79–88. doi:10.1016/0360-3199(82)90209-9
49. Ambriz JJ, Ducarroir M, Sibieude F (1982) Preparation of cadmium by thermal dissociation of cadmium oxide using solar energy Int J Hydrogen Energy 7:143–153. doi:10.1016/0360-3199(82)90141-0
50. Weidenkaff A, Steinfeld A, Wokaun A, Auer PO, Eichler B, Reller A (1999) Direct solar thermal dissociation of zinc oxide: condensation and crystallisation of zinc in the presence of oxygen. Sol Energy 65:59–69. doi:10.1016/S0038-092X(98)00088-7
51. Lundberg M (1993) Model calculations on some feasible two-step water splitting processes. Int J Hydrogen Energy 18:369–376. doi:10.1016/0360-3199(93)90214-U
52. O'Keefe D, Allen C, Besenbruch G, Brown L, Norman J, Sharp R (1982) Preliminary results from bench-scale testing of a sulfur-iodine thermochemical water-splitting cycle. Int J Hydrogen Energy 7:381–392. doi :10.1016/0360-3199(82)90048-9
53. Sakurai M, Nakajima H, Amir R, Onuki K, Shimizu S (2000) Experimental study on side-reaction occurrence condition in the iodine-sulfur thermochemical hydrogen production process. Int J Hydrogen Energy 25:613–619. doi:10.1016/S0360-3199(99)00074-9
54. Kubo S, Nakajima H, Kasahara S, Higashi S, Masaki T, Abe II (2004) A demonstration study on a closed-cycle hydrogen production by the thermochemical water-splitting iodine sulfur process. Nucl Eng Des 233:347–354. doi:10.1016/j.nucengdes.2004.08.025
55. Kameyama H, Yoshida K (1981) Reactor design for the UT-3 thermochemical hydrogen production process. Int J Hydrogen Energy 6:567–575. doi:10.1016/0360-3199(81)90022-7
56. Kameyama H, Tomino Y, Sato T, Amir R, Orihara A, Aihara M (1989) Process simulation of "Mascot" plant using the UT-3 thermochemical cycle for hydrogen production. Int J Hydrogen Energy 14:323–330. doi:10.1016/0360-3199(89)90133-X
57. Sakurai M, Bilgen E, Tsutsumi A, Yoshida K (1996) Adiabatic UT-3 thermochemical process for hydrogen production. Int J Hydrogen Energy 21:865–870. doi:10.1016/0360-3199(96)00024-9
58. Miyoka H, Ichikawa T, Nakamura N, Kojima Y (2012) Low temperature water splitting by sodium redox reaction. Int J Hydrogen Energy 37:17709–17714. doi:10.1016/j.ijhydene.2012.09.085
59. Weimer A (2008) H2A analysis for Manganese oxide based solar thermal water splitting cycle. University of Colorado STCH, Denver
60. Tamura Y, Steinfeld A, Kuhn P, Ehrensberger K (1995) Production of solar hydrogen by a novel, 2-step, water-splitting thermochemical cycle. Energy 20:325–330. doi:10.1016/0360-5442(94)00090-O

61. Sturzenegger M, Ganz J, Nüesch P, Schelling T (1999) Solar hydrogen from a manganese oxide based thermochemical cycle. J de Phys Arch 09:Pr3–331–Pr3–335. doi:10.1051/jp4:1999351

62. Xu B, Bhawe Y, Davis ME (2012) Low-temperature, manganese oxide-based, thermochemical water splitting cycle. Proc Natl Acad Sci 109:9260–9264. doi:10.1073/pnas.1206407109

63. Leitner W, Dinjus E, Gaßner F (1994) Activation of carbon dioxide: IV. Rhodium-catalysed hydrogenation of carbon dioxide to formic acid. J Organomet Chem 475:257–266. doi:10.1016/0022-328X(94)84030-X

64. Coffey RS (1967) The decomposition of formic acid catalysed by soluble metal complexes. Chem Commun 18:923b–924. doi:10.1039/C1967000923B

65. Yoshida T, Ueda Y, Otsuka S (1978) Activation of water molecule. 1. Intermediates bearing on the water gas shift reaction catalyzed by platinum (0) complexes. J Am Chem Soc 100:3941–3942. doi:10.1021/ja00480a054

66. Paonessa RS, Trogler WC (1982) Solvent-dependent reactions of carbon dioxide with a platinum (II) Dihydride reversible formation of a platinum(II) formatohydride and a cationic platinum(II) dimer, [Pt2H3(PEt3)4][HCO2]. J Am Chem Soc 104:3529–3530. doi:10.1021/ja00376a058

67. Joszai I, Joo F (2004) Hydrogenation of aqueous mixtures of calcium carbonate and carbon dioxide using a water-soluble rhodium(I)–tertiary phosphine complex catalyst. J Mol Catal A: Chem 224:87–91. doi:10.1016/j.molcata.2004.08.045

68. Gao Y, Kuncheria JK, Yap GPA, Puddephatt RJ (1998) An efficient binuclear catalyst for decomposition of formic acid. Chem Commun 21:2365–2366. doi:10.1039/A805789C

69. Shin JH, Churchill DG, Parkin G (2002) Carbonyl abstraction reactions of Cp*Mo(PMe$_3$)$_3$H with CO$_2$, (CH$_2$O)$_n$, HCO$_2$H, and MeOH: the synthesis of Cp*Mo(PMe$_3$)$_2$(CO)H and the catalytic decarboxylation of formic acid. J Organomet Chem 642:9–15. doi:10.1016/S0022-328X(01)01218-9

70. Loges B, Boddien A, Gartner F, Junge H, Beller M (2010) Catalytic generation of hydrogen from formic acid and its derivatives: useful hydrogen storage materials. Top Catal 53:902–914. doi:10.1007/s11244-010-9522-8

71. Fukuzumi S, Suenobu T, Ogo S. Catalysts for the decomposition of formic acid, method for decomposing formic acid, process for producing hydrogen, apparatus for producing and decomposing formic acid, and method for storing and producing hydrogen, US Patent application no. US2010/ 0034733 A1

72. Guo WL, Li L, Li LL, Tian S, Liu SL, Wu YP (2011) Hydrogen production via electrolysis of aqueous formic acid solutions. Int J Hydrogen Energy 36:9415–9419. doi:10.1016/j.ijhydene.2011.04.127

73. Majewski A, Morris DJ, Kendall K, Wills M (2010) A continuous-flow method for the generation of hydrogen from formic acid. ChemSusChem 3:431–434. doi:10.1002/cssc.201000017

74. Loew O (1887) Ueber einige katalytische Wirkungen. Berichte der deutschen chemischen Gesellscha 20:144–145

75. Kapoor S, Naumov S (2004) On the origin of hydrogen in the formaldehyde reaction in alkaline solution. Chem Phys Lett 387:322–326. doi:10.1016/j.cplett.2004.01.127

76. Ansell MF, Coffey S, Rodd EH (1965) Rodd's chemistry of carbon compounds Elsevier. Page, Amsterdam 11

77. Ashby EC, Doctorovich F, Liotta CL, Neumann HM, Barefield EK, Konda A, Zhang K, Hurley J, Siemer DD (1993) Concerning the formation of hydrogen in nuclear waste. Quantitative generation of hydrogen via a Cannizzaro intermediate. J Am Chem Soc 115:1171–1173. doi:10.1021/ja00056a065

78. Harden A (1899) Formaldehyde, action of hydrogen peroxide on. J Soc Chem Indus 15:158–159

79. Satterfield CN, Wilson RE, Le Clair RM, Reid RC (1954) Analysis of aqueous mixtures of hydrogen peroxide and aldehydes. Anal Chem 26:1792–1797. doi:10.1021/ac60095a030

80. Gorse RA, Volman DH (1971) Analysis of mixtures of hydrogen peroxide and formaldehyde. Anal Chem 43:284–284. doi:10.1021/ac60297a031
81. Kurt C, Bittner J (2006) Sodium hydroxide: Ullman's encyclopedia of industrial chemistry. doi:10.1002/14356007.a24_345.pub2
82. Kumar S (2013) Clean hydrogen production and carbon dioxide capture methods. FIU electronic theses and dissertations paper 1039. http://digitalcommons.fiu.edu/etd/1039
83. Kreider PB, Funke HH, Cuche K, Schmidt M, Steinfeld A, Weimer AW (2011) Manganese oxide based thermochemical hydrogen production cycle. Int J Hydrogen Energy 36:7028–7037. doi:10.1016/j.ijhydene.2011.03.003
84. Charvin P, Abanades S, Beche E, Lemont F, Flamant G (2009) Hydrogen production from mixed cerium oxides via three-step water-splitting cycles. Solid State Ionics 180:1003–1010. doi:10.1016/j.ssi.2009.03.015

Chapter 3
Modified Steam Methane Reformation Methods for Hydrogen Production

Abstract Over the past few decades, extensive efforts have been made to modify the conventional hydrogen production technologies. In particular, a review of those methods that are intended to reduce the carbon emission and improve the process efficiency for steam methane reformation (SMR) is provided here. So far, several such methods have been proposed based on both fossil and non-fossil energy sources which primarily include the use of membranes, metal oxides as a CO_2 sorbent, and nuclear and solar energy. Moreover, this section also includes a brief summary of an innovative process which suggests the inclusion of sodium hydroxide as a reactant to the SMR process. The addition of sodium hydroxide to the SMR process can serve the dual purpose of hydrogen production and CO_2 capture. Certainly, these methods have the potential to reduce CO_2 emission during hydrogen production. Therefore, here, the status and perspective of all these methods are presented.

Keywords Carbon emission · SMR · Membrane · CO_2 sorbent

3.1 Introduction

Currently, global hydrogen production is dominated by the steam methane reformation (SMR). Since the establishment of the SMR process in 1930, intensive research and development in the field has been responsible for higher catalyst performance and better reactor tube materials [1]. The efforts have been made continuously to

(i) improve the catalyst performance to provide enhanced activity, higher mechanical strength, better resistance to carbon formation and sulfur poisoning [2–4], and

(ii) also improve the ability of the reactor tube material to withstand higher stresses at elevated temperature and thermal flux [5–9].

Since about 1970, another research area emerged related to the reforming reactor configuration and the suitability of the multi-tube fixed bed reactor for the

© The Author(s) 2015
S. Kumar, *Clean Hydrogen Production Methods*,
SpringerBriefs in Energy, DOI 10.1007/978-3-319-14087-2_3

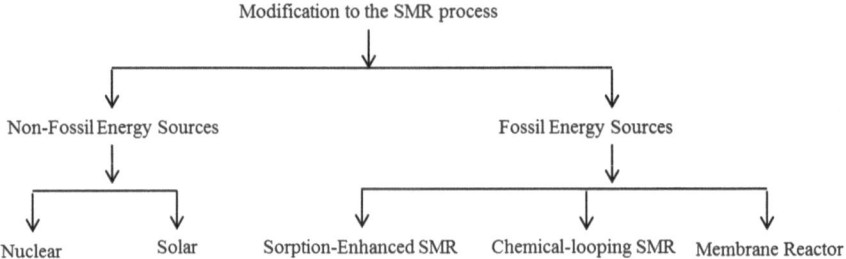

Fig. 3.1 Modifications to the SMR technique

reforming reactions. To date, numerous attempts have been made to radically improve the performance of reforming process through changes in reactor configuration. The improvements that have been proposed are as follows: (i) changing from a fixed bed to fluidized bed reactor, (ii) changing from external firing to direct heating, and (iii) utilizing membrane technology to drive the reaction beyond the thermodynamic equilibrium conversion [10–12]. A thorough analysis of these areas of improvement and the suitability of proposed configurations for different reforming applications can be found elsewhere [13].

SMR is an energy-intensive process that adds a significant amount of carbon dioxide per unit of hydrogen produced. Therefore, in 1988, Brun–Tsekhovoi et al. [14] proposed a process for catalytic steam reforming of hydrocarbons in the presence of CO_2 acceptors. Here, the aim is to list all similar methods related to SMR technique. Figure 3.1 summarizes such schemes and categorizes them on the basis of the type of energy input. A detailed explanation of each proposed technique is provided in the following sections.

3.2 Steam Methane Reforming Using Non-fossil Energy Sources

The use of alternative energy sources, such as high-temperature nuclear and solar heat for hydrogen production, has gained substantial interest of researchers worldwide. Besides being energy efficient, these alternative energy sources have the capacity to significantly mitigate carbon emission. However, one of the foremost challenges is to develop these techniques at a large scale. A short description for both high-temperature nuclear and solar heat methods is as follows:

3.2.1 Nuclear Energy

Nuclear power systems, a relatively clean and abundant energy source, have the potential to contribute to hydrogen economy. Nuclear energy can be used to generate both electricity and hydrogen. The production of hydrogen through nuclear energy is economically sound but would require major technological development in the coming years [15]. Moreover, the future goal should be to change the public perception about nuclear energy in a manner that significantly contributes to the hydrogen economy.

Nuclear energy can produce hydrogen via water electrolysis, thermochemical splitting, or a hybrid process. To facilitate these hydrogen production technologies, nuclear plants can provide both heat and electricity. There are three ways to utilize the nuclear energy for hydrogen generation:

(i)   electricity for electrolysis of water,
(ii) both heat and electricity for high-temperature steam electrolysis or hybrid process, and
(iii) heat to run thermochemical processes (generates hydrogen from hydrocarbons or water).

As thermochemical processes are very energy intensive, nuclear energy can be seen as a solution. Thus, here, a short description for coupling nuclear reactor with the SMR process is given.

The generation of hydrogen through nuclear energy has important advantages over other processes. For instance, it does not require fossil fuels, which results in lower greenhouse gas emissions, and can lend itself to a large-scale production. The hydrogen production properties determine the type of reactors that can suitably be coupled to the hydrogen production scheme. The design for the electrochemical and thermochemical hydrogen production technology should consider the following requirements:

(i)   high-temperature for achieving high thermal to hydrogen efficiency,
(ii) high thermal to electrical power conversion efficiency,
(iii) minimum temperature loss for the reactor coolant in order to achieve effective heat transfer rate to the chemical plant,
(iv) minimum pressure losses in the primary loop,
(v)   selection of chemically inert coolants to ensure high safety, and
(vi) low capital costs.

Since high temperature is required for the thermochemical or electrochemical process of hydrogen production, gas-cooled reactors, molten-salt-cooled reactors, and heavy-metal-cooled reactors, all coupled with gas power cycles, appear to be the most promising technologies for the hydrogen production [16].

A modular helium nuclear reactor (MHR) can substitute for the natural gas burning furnaces as a heat source for the SMR process. As the MHR can operate at about 850 °C, the efficiency of the process is about 80 %. The concept to integrate MHR to the SMR process can be potentially cost competitive to the conventional SMR. Also, the MHR-SMR can significantly reduce CO_2 emissions.

In MHRs, recycled helium is heated to 850 °C, which is suitable for the SMR reaction. The operating pressure of the MHR is 70 bars. Hot helium flows inside an indirectly heated heat exchanger countercurrent to methane and steam. In the

Table 3.1 Overview of nuclear hydrogen production technologies using thermochemical SMR process [20]

Feature	SMR
Reaction	Reforming: $CH_4 + H_2O \rightarrow CO + 3H_2$, endothermic (750–800 °C) Shift: $CO + H_2O \rightarrow CO_2 + H_2$, exothermic (350 °C)
Temperature (°C)	>700
Efficiency (%)	>60, Temperature dependent
Energy efficiency coupled to MHR (%)	>60, Temperature dependent
Advantage	Mitigate CO_2 emission
Disadvantage	Depends on methane prices

process, helium releases its sensible heat to methane steam streams through the reformer tube and eventually cools down to 600 °C [17]. The reformer tube has an inner helical tube through which heat is dissipated to the catalyst-filled tube. Helium heat carrier present in the core of a high-temperature nuclear reactor is heated to run the process. Such reactor has been examined at a pilot scale and is anticipated to be commercialized for syngas production, thermochemical sulfur–iodine process [18], or the Westinghouse sulfur process [19]. The system can incorporate electricity generation equipment to meet cogeneration needs. Moreover, the reformed gas can be used to produce valuable chemicals (such as H_2, NH_3, and CH_3OH) or transfer heat to long distance (in conjunction with methanation) (Table 3.1).

The key challenge is to couple the nuclear reactor system with steam methane reforming process. In 2004, Japan Atomic Energy Agency (JAEA) demonstrated a nuclear-heat-driven small-scale SMR-based hydrogen production unit. It can be easily understood that the economy of nuclear steam reforming strongly depends on the cost of natural gas [21].

3.2.2 Solar Energy

Similarly, attempts have been made to harness solar energy. There are basically three ways of producing hydrogen with the aid of solar energy: electrochemically, photochemically, and thermochemically. At present, solar concentrators can provide solar flux concentrations in three ranges: trough concentrators (30–100 suns), tower systems (500–5,000 suns), dish systems (1,000–10,000 suns). For a solar concentration of 5,000, the optimum temperature of the solar receiver is about 1,270 °C, attaining a maximum theoretical efficiency of 75 %; a theoretical efficiency is defined as the portion of solar energy that can be converted to chemical energy of fuels [22]. Such a high temperature is sufficient to conduct energy-intensive methods such as SMR and CO_2 reforming of methane. One should note that the solar chemical reactors are equipped with a well-insulated enclosure with a small opening (the aperture) that allows the solar radiation to penetrate in.

The solar energy can be used via two processes: (a) open loop and (b) closed loop. Figure 3.2 shows the schematic of both process options. In the open-loop system, the hydrocarbon feedstock uses solar energy to produce on-site syngas for subsequent combustion in a conventional gas turbine or a combined cycle power plant. Notably, this route is energetically more favorable compared to simply producing steam using the solar energy because it harvests the solar energy in a chemical form, rather than as sensible heat. Syngas can be directed to fuel cells or be utilized for the formation of specialty chemicals and plastics and liquid fuels (methanol and gasoline).

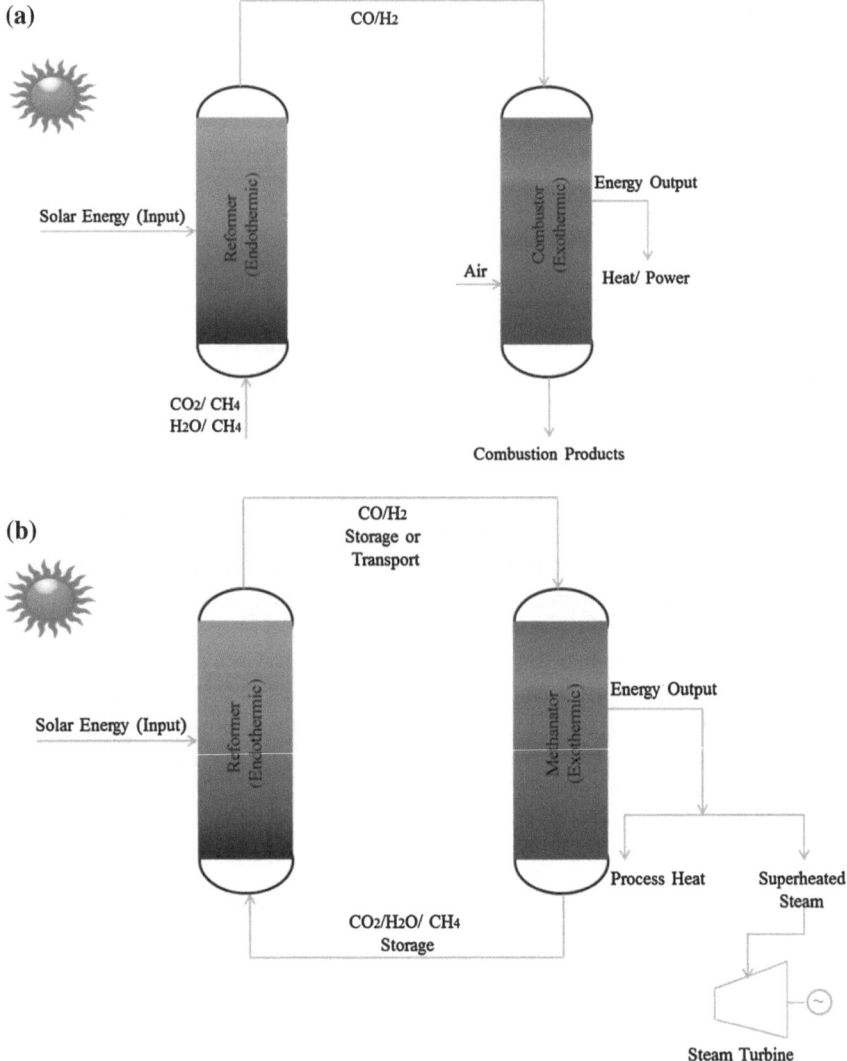

Fig. 3.2 Concepts for **a** open- and **b** closed-loop thermochemical heat pipes based on CH_4/CO_2 reforming and solar energy adapted from [23]

Figure 3.2b shows the closed loop, where methane feedstock is converted to syngas using solar energy. The syngas can then be either stored or transported off-site prior to conversion to methane in a methanation reactor that recovers solar energy in the form of heat to run industrial processes or generate electricity. Both methanation reactions ($CO + 3H_2 \rightarrow CH_4 + H_2O$ and $2CO + 2H_2 \rightarrow CH_4 + CO_2$) are exothermic in nature and in principle and can be integrated with a conventional steam turbine power plant or be used for the production of high-grade process heat. The heat should be extracted in a controlled way, and thus, maintaining the temperature is a critical task for exothermic methanation reactions. Conventionally, the temperature control was performed by using a high recycle and thus diluting the inlet gas in order to keep the temperature below 450 °C. Currently, there are various ways to recover the heat as high-pressure supersaturated steam at 100 bar/540 °C, which can be sent directly in a steam turbine power cycle [24].

Solar reforming of methane is performed in solar furnaces as well as in solar simulators using different reactor configurations and catalysts. Solar chemical reactor for highly concentrated solar system uses a well-insulated enclosure with a small aperture to allow in concentrated solar radiation. Aristov et al. [25] conducted experiments for steam reforming of methane under direct illumination of a catalyst by concentrated light in the energy receiver with a transparent wall. Such reactor showed a considerable increase in both the specific rate of hydrogen production and the specific power loading of the light-to-chemical energy conversion compared to a conventional stainless steel reactor. Precisely, 1 g of catalyst generated 130 Ndm^3/h of hydrogen and a power of 50–100 W/cm^3. The increase in the reaction rate is attributed to the direct absorption of light by catalyst granules, which intensified the energy input to the catalyst bed. Yokota et al. [26] used xenon lamp (a solar simulator) to conduct steam reforming of methane over Ni/Al_2O_3 catalyst. The ratio of $H_2O/CH_4 = 1/1$ was fixed, but the temperature varied in the range of 650–950 °C. At 850 °C, methane conversion was in excess of 85 % under atmospheric pressure.

The reaction between methane and steam is preferred if hydrogen is the desired product. However, if methanol is desired as end product, then CO_2 reforming of methane is chosen. Recently, a new configuration has been tested which involves the mixed reforming (both steam and CO_2) for methane. The process is particularly advantageous for biogas, with a high CO_2 content (45–70 mol% CH_4 and 30–45 mol% CO_2) [27, 28]. In summary, considerable efforts have been made recently for coupling renewable solar energy and conventional fossil fuels to synthesize hydrogen or valuable chemicals.

3.3 Steam Methane Reforming Using Fossil Energy Sources

In this section, a review of the modified SMR technique based on the use of fossil fuels is provided. Most of these proposed methods include metal oxides, perovskites, alkali, and membrane that can result in less carbon emission compared to the SMR method. As can be understood, such methods cannot completely eliminate the carbon emission via the SMR process but would mitigate these emissions to a large extent. Considering the availability of fossil fuels for a definite period, these methods currently have a role in satisfying the ever-increasing global energy demand. In the following paragraphs, these methods are presented with their relevant experimental results.

3.3.1 Sorption-Enhanced Steam Methane Reforming (SE-SMR)

Sorbent-enhanced steam methane reforming (SE-SMR) is a technology for the production of high-purity hydrogen from hydrocarbons with in situ CO_2 capture. Figure 3.3 depicts the sorption-enhanced reforming scheme. In the SE-SMR process, hydrocarbon reforming, Eq. (3.1), water gas shift, Eq. (3.2), and CO_2 separation reactions Eq. (3.3) occur simultaneously in a single reaction step over a reforming catalyst mixed with a CO_2 sorbent. The overall reaction is given in Eq. (3.4).

$$CH_4(g) + H_2O(g) \leftrightarrow CO(g) + 3H_2(g) \tag{3.1}$$

$$CO(g) + H_2O(g) \leftrightarrow CO_2(g) + H_2(g) \tag{3.2}$$

$$MO(s) + CO_2(g) \leftrightarrow MCO_3(s) \tag{3.3}$$

$$CH_4(g) + 2H_2O(g) + MO(s) \leftrightarrow MCO_3(s) + 4H_2(g) \tag{3.4}$$

MO denotes metal oxide that transforms to carbonate (MCO_3) after reaction with CO_2. An ideal sorbent should lead to high yields of hydrogen and negligible concentrations of CO, CO_2, and unreacted methane. As can be seen that the hydrocarbon reforming is endothermic but by inclusion of a carbonate-forming sorbent, the

Fig. 3.3 Schematic for sorption-enhanced reforming of methane

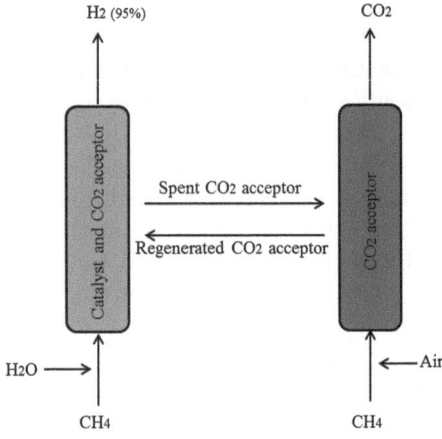

overall reaction turns out to be thermoneutral. However, the sorbent regeneration step requires substantial energy input. The SE-SMR method has been proposed to (i) avoid the operational complexity of the SMR technique and (ii) alleviate the high costs associated with the amine-based CO_2 capture technology. The advantages of the SE-SMR process are as follows: (1) fewer processing steps; (2) improved energy efficiency; (3) elimination of the need for shift catalysts; and (4) reduction in the temperature of the primary reforming reactor by 150–200 °C [29, 30].

It has been realized that the integration of heat between the exothermic sorption reaction and the endothermic reforming reaction can improve the energy efficiency. Since both the SE-SMR process and the sorbent regeneration are solid–gas methods, a fixed bed or fluidized bed reactors can be employed [31–34]. Lindborg and Jakobsen [35] mentioned that the circulating fluidized bed reactor may have better advantages compared to fixed bed reactors regarding heat integration and continuous operation for both the sorption of CO_2 and the regeneration of the sorbent. To date, an extensive amount of experimental as well as modeling work has been performed using the fixed bed reactor for the SE-SMR process. Wang et al. [36] have not only listed most of the previous work based on fixed bed reactors but also simulated a circulating fluidized bed reactor operating in a continuous mode within two sets of 3D cylindrical coordinates for the downer and riser. Interestingly, Yi and Harrison [37] obtained high H_2 and very low CO concentrations in their experiments for SE-SMR process conducted at relatively low reaction temperatures (400–460 °C) and low pressure (1–5 bar) over H_2O–CH_4 feed ratio (3:1) in a laboratory-scale fixed bed reactor. Similarly, Balasubramanian et al. [32] showed that the combination of sorption and reforming reaction was significantly fast and the equilibrium was almost attained under all the reaction conditions performed [temperature (450–750 °C), CH_4 to feed gas (6–20 mol%), steam-to-methane ratio (3–5), total gas feed rate (200–1,200 cm^3 STP/min), reforming catalyst (3–7 g), and CaO (7–10 g)]. The selection of suitable reactor type primarily depends on the capacity of the sorbent and the rates of adsorption and desorption. A fixed bed reactor is mostly suited for high capacity and low reaction rate, and the reverse is true for fluidized bed reactor.

Different CO_2 sorbents, such as CaO, dolomite [38, 39], lithium zirconate (Li_2ZrO_3) [40], sodium zirconate (Na_2ZrO_3) [41], lithium silicate (Li_4SiO_4) [42], or hydrotalcite [43, 44], have been assessed for the SE-SMR. A detailed review which presents a comparative study of these sorbents with CO_2 can be found elsewhere [45]. These sorbents should (i) lock enough CO_2 in the temperature range of 400–600 °C; (ii) withstand $P_{H2O}/P_{CO2} > 20$ at a steam reforming condition; (iii) be stable both mechanically and chemically for long hours at such high-temperature and high-pressure conditions; and (iv) have fast adsorption and desorption kinetics.

Ca-based sorbents are more widely investigated and used because of their high CO_2 capacity and rapid reaction rate, as well as lower cost [46, 47]. Currently, most of the efforts are dedicated to improve the multi-cycle durability of CaO [48]. The capture of CO_2 using CaO is based on the well-established carbonation–calcination cycle. CaO can reversibly lock CO_2 in the form of calcium carbonates ($CaCO_3$) (reaction 3.6).

$$CH_4(g) + CaO(s) + 2H_2O(g) \rightarrow CaCO_3(s) + 4H_2(g) \quad \Delta H^\circ_{298K} = -13kJ/mol$$

$$(3.5)$$

$$CaCO_3(s) \rightarrow CaO(s) + CO_2(g) \quad \Delta H^\circ_{298K} = +178.1 \ kJ/mol \qquad (3.6)$$

The heat released by the exothermic carbonation reaction (CaO(s) + CO$_2$(g) $\rightarrow$ CaCO$_3$(s), ΔH°_{298K} = -178.1 kJ/mol) balances the heat input required by the endothermic reforming reaction (a reverse of the carbonation reaction). Therefore, additional fuel may not be required in the reforming reactor. Recent work of Ochoa–Fernández et al. [46] illustrated that in spite of high regeneration energy demand, CaO-based SE-SMR processes can attain thermal efficiencies as high as 82 % compared to 71 % for a conventional process scheme based on amine solvents.

Figure 3.4 depicts the thermodynamic equilibrium calculation for the CH$_4$–H$_2$O–CaO system using the software FACTSAGE^{TM} and the databases therein. The thermodynamic calculation clearly shows that (i) the modified reaction captures CO$_2$ in the form of CaCO$_3$ (stable at a high temperature); (ii) hydrogen can be produced at relatively lower temperatures in the proposed system; and (iii) about 88 % methane can be converted, and that would result in 95 vol.% H$_2$. The regeneration of the spent CaO can be accomplished in the adiabatic fluidized bed reactor generator at about 975 °C. Once regenerated, CaO can be fed back to the reforming reactor.

Several issues that limit the application of CaO for the SE-SMR at a large scale are as follows:

(i)   high regeneration temperature of CaO (>975 °C) leads to vast energy penalty;
(ii) continuous separation of reforming catalyst and the CO$_2$ acceptor (CaO); and
(iii) need to improve the durability of CO$_2$ acceptor for conducting multiple-cycle operations

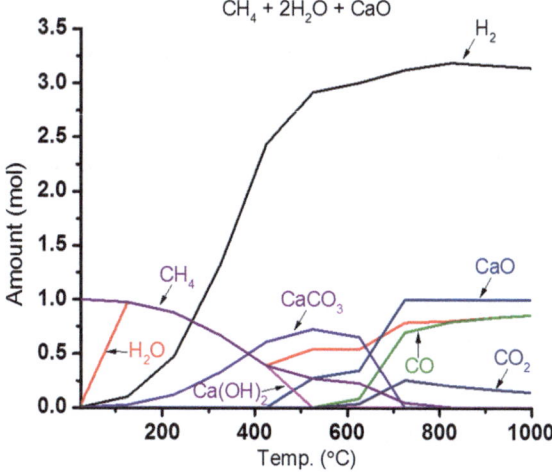

Fig. 3.4 Equilibrium in the CH$_4$–H$_2$O–CaO system

CaO-derived calcination of naturally occurring limestone has poor cyclic CO_2 capture stability. The decrease in CO_2 uptake of limestone with increasing number of calcination and carbonation cycles has been attributed to thermal sintering resulting in the loss of pore volume within pores of diameter <100 nm [49, 50]. So far, several methods have been proposed to improve the CO_2 capture properties of limestone that include the thermal pretreatment or hydration [51–53]. Also, CaO is stabilized using high-Tammann-temperature support such as Al_2O_3, ZrO_2, or MgO, to achieve a sinter-resistant material [54–60]. Recent finding suggests that hydrogen concentration in the product stream was higher for Ni/CaO compared to Ni/Al_2O_3. At 600 °C, 1 atm and steam-to-methane ratio of 3, 12.5 wt% of Ni/CaO has produced 80 % hydrogen concentration. This result illustrates that multi-functional catalysts can be used as a better alternative to Al_2O_3 and can also reduce the reactor size [61]. Moreover, to achieve the slower capacity degradation of CaO-based sorbent, these sorbents have been doped with inert materials such as $KMnO_4$ [62], $CaTiO_3$ [63], $Ca_{12}Al_{14}O_{33}$ [64, 65], and $Ca_9Al_6O_{18}$ [66].

Owing to the existing issues related to CaO, recent efforts have been aimed to modify CaO-based sorbents or employed synthetic sorbents. Among synthetic sorbents, layered double hydroxides (LDH) are extensively studied due to their high surface area, low regeneration temperature, faster kinetics compared to CaO, stable framework, slow degradation in sorption capacity, and presence of abundant basic sites (favorable for reaction with acidic CO_2) [67–71]. However, the main issue with LDH is the low CO_2 capture capacities of about 1 mmol/g. Extensive work has been carried out to improve the CO_2 capture capacities of LDH which includes the doping of LDH with alkali metal carbonates, precipitation on different support to improve the LDH surface area, and the surface modification [66]. Despite considerable efforts, no significant improvement related to the CO_2 capture capacity of LDH has been reported so far.

To date, sorbents have been primarily assessed for their multi-cyclic properties to capture CO_2, and thus, relatively little information is available on their application in the overall sorption-enhanced hydrogen production process. Due to less favorable thermodynamic properties of these CO_2 acceptors, the equilibrium CO_2 pressures are generally high. Moreover, slow kinetics of the CO_2 sorption impedes the use of SE-SMR for hydrogen production on a large scale.

3.3.2 Chemical Looping Steam Methane Reforming (CL-SMR)

Chemical looping combustion (CLC) of fossil fuel has not only an inherent advantage of nearly zero energy consumption in capturing CO_2, but also has the ability to generate pure hydrogen [72, 73]. Thus, numerous researchers have studied CLC systems that focus mainly on fossil fuels [74–76]. In this method, a reducing agent is employed to reduce metal oxide, and then, an H_2O oxidation step is performed to reoxidise the metal and produce hydrogen [77]. A CLC configuration comprises of two reactors: (i) reduction reactor (reduces fuels using metal oxide as an oxygen carrier) and (ii) oxidation reactor [air oxidation of reduced oxygen carrier (N_2 is obtained as the product gas)]. CLC system has high thermal energy efficiency and has application in thermal power generation. In general, CLC uses fluidized bed reactor and works at above 900 °C [78, 79]. However, sintering and fast degradation of oxygen carriers as well as carbon deposition have emerged as issues to overcome [80, 81].

In order to produce clean hydrogen, Ryden et al. [82] proposed a model that integrates both CLC and SMR. The reactor system consists of two interconnected fluidized beds. The oxidation reactor is a high-velocity fluidized bed acting as a riser, while the reduction reactor is low-velocity bubbling bed. Metal oxide can be used as bed material. The high gas velocity in the oxidation reactor ensures the continuous circulation of particles between the beds. Oxidized particles are collected in the cyclone and fed to the reduction reactor. Reduced oxygen carrier is transferred back to the oxidation reactor with the aid of gravity. As can be expected, heat for endothermic reforming reaction is provided by the exothermic reaction in the oxidation reactor and transferred to the reduction reactor with the particle circulation. Few advantages could be listed for such system: no formation of thermal NOx, H_2-rich off-gas, hot fluidized particles maintains the temperature of reformer tube walls, relatively easy reactor design as CLC takes place at ambient pressure and modest temperature, and off-gas from the pressure swing adsorption unit is used as fuel in the reduction reactor.

CLC uses metal oxides to circulate oxygen (Fig. 3.5). Thus, a suitable metal oxide needs to be identified. So far, NiO has gained the maximum attention as it has excellent reactivity, significant thermal stability, high oxygen-carrying capacity, and low volatilization at a high temperature [83–85]. However, NiO is expensive and toxic [86]. In the same line, Fe-oxides have been extensively studied owing to its low cost, high melting point, and almost zero carbon deposition [87–89]. Another vital issue is Fe_2O_3 cannot be directly reduced to Fe, but only to FeO [90]. Oritz et al. [91] evaluated an iron-based waste as oxygen carrier obtained from aluminum manufacture called "sand process." This material showed enough high oxygen transport capacity (2.4 wt%) and reactivity to be able to convert a simultaneous syngas to CO_2 and H_2O at 880 °C. However, lower conversion of fuel was noticed for methane-containing fuels. Cho et al. [92] used 20 wt% Fe_2O_3/ZrO_2 as oxygen carrier for SR-CLC process. The results illustrated the average steam conversion is 35 %, but decreases sharply in the later oxidation zone. Another study showed an improvement in the redox reactivity of SR-CLC process by using $Ce_{1-x}Fe_xO_{2-\delta}$ [93]. However, overall, the reactivity of Fe-oxides is inferior to Ni- and Cu-based oxygen carriers [94, 95].

Interestingly, Cu shows high reactivity as well as significant oxygen-carrying capacity and also does not react with inert components [96]. Moreover, the reduction of CuO and oxidation of Cu are both exothermic in nature. Thus, CuO does not require heat to maintain the working temperature for reduction. However, Cu-based oxygen carriers suffer from the sintering problem as melting point of metallic Cu is relatively low, 1,083.4 °C [97]. Hence, in order to use Cu-based

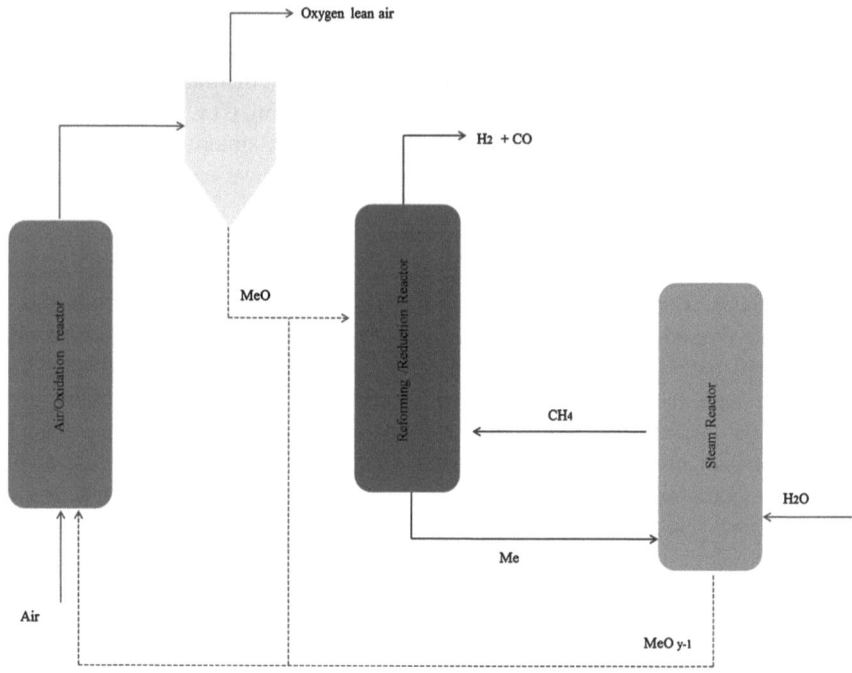

Fig. 3.5 Schematic of SMR-CL system for syngas and hydrogen production

oxygen carriers, one needs to reduce the reaction temperature. Zheng et al. [98] prepared CuO–SiO_2 oxygen carrier through wet impregnation method and illustrated the reduction in the CLC working temperature by transforming CH_4 to H_2 and CO. Other oxides such as Mn_2O_3 [99, 100], CoO [101], and $CaSO_4$ [102, 103] have also been considered as potential oxygen carriers. A recent study considered bimetallic carriers based on iron and manganese (Mn_xFe_{1-x}–CeO_2) supported on CeO_2 as oxygen carriers in CLC. The results showed that the use of ceria as support resulted in stable operation. Fe-rich carriers exhibited an unusual reversible dealloying/realloying behavior during cyclic redox reactions. However, Mn-rich carriers showed a noticeable increase in carrier reactivity and selectivity for total oxidation of methane [104].

In 2014, Zhao et al. [105] reported the synthesis of a three-dimensionally ordered macroporous (3DOM) $LaFeO_3$ and nano-$LaFeO_3$ perovskites which can be used as oxygen carriers for SR-CLC process. The performance of the perovskites as oxygen carriers to produce syngas and hydrogen was investigated. The net reactions of SR-CLC can be illustrated as follows:

$$\text{Methane reduction:} \, LaFeO_3 + H_2O \rightarrow LaFeO_{3-(\delta 1 + \delta 2)} + (CO + H_2) \qquad (3.7)$$

$$\text{Steam oxidation: LaFeO}_{3-(\delta_1+\delta_2)} + H_2O \rightarrow \text{LaFeO}_{3-\delta_1} + H_2 \qquad (3.8)$$

$$\text{Air oxidation: LaFeO}_{3-\delta_1} + O_2 \rightarrow \text{LaFeO}_3 \qquad (3.9)$$

The results showed that 3DOM $LaFeO_3$ perovskites have good repeatability, more stable reactivity of methane oxidation, and better resistance to carbon formation than nano-$LaFeO_3$. The better reactivity of 3DOM $LaFeO_3$ was attributed to its higher surface area. Another set of studies were conducted for perovskites ($La_{0.6}Sr_{0.04}Co_{0.02}Fe_{0.8}O_{3-\delta}$ and $La_{0.7}Sr_{0.3}FeO_{3-\delta}$) against two supported metal oxides (60 % Fe_2O_3/Al_2O_3 and 20 % NiO/Al_2O_3). $La_{0.7}Sr_{0.3}FeO_{3-\delta}$ and 60 % Fe_2O_3/Al_2O_3 exhibited better performance than other combinations. However, due to the formation of $FeAl_2O_4$, the performance for 60 % Fe_2O_3/Al_2O_3 was downgraded for large number of cycles (~ 150) [106].

The aforementioned limitations of metal oxides (CuO, NiO, and Fe_2O_3) have led to grow interest for perovskites materials, which so far seems very interesting. It is certain that there is a need to develop SR-CLC process for the generation of clean hydrogen. However, an efficient oxygen carrier with sufficient carrier stability and fast redox kinetics is still to be explored. The continuous cycling of particles between reduction and oxidation reactor subjects the oxygen carrier to chemical, thermal, and mechanical stress. Therefore, efforts need to be focused to develop efficient oxygen carriers that can cope with such harsh environment.

3.3.3 Hydrogen Membrane Reactor

The concept of a hydrogen membrane reactor is based on Le Chatelier's principle according to which the hydrogen produced in the reactor permeates through a membrane and exits through the reaction zone (Fig. 3.6). Membranes are several

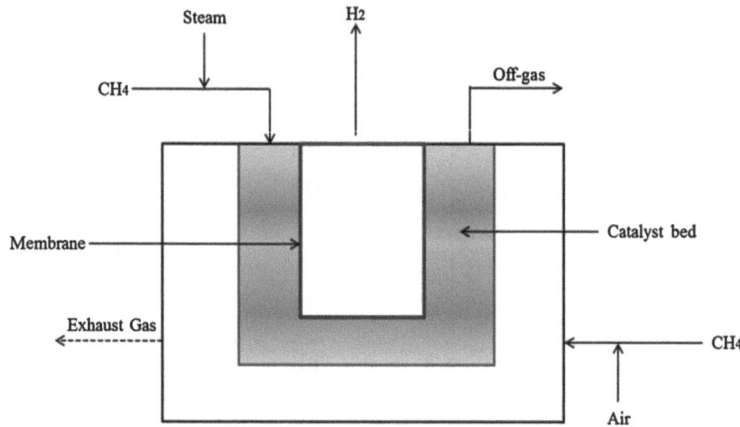

Fig. 3.6 Schematic of hydrogen membrane reactor

microns thick and generally made of Pd [107] or Pd/Ag [108] or other Pd-based alloys [109]. Pd is the most used membrane material for hydrogen permeation because it is infinitely selective to hydrogen and has shown to have high permeability. Pd is usually alloyed with metal to increase permeability and reduce the effect of hydrogen embrittlement [110].

The main advantages of the membrane technique are as follows: (a) no limitations of chemical equilibrium for the hydrogen-producing reaction, which means the equilibrium is shifted more toward product, (b) higher conversion of methane can be achieved at lower temperatures, (c) generation of separate H_2 and CO_2 streams, (d) no need of additional CO-shift converters, (e) more simple and compact reactor configurations, and (f) high overall efficiency owing to the relatively low temperature resistance of the Pd-based membranes; hydrogen membrane reactors operate at temperatures of 400–600 °C (compared to 800–950 °C typical of conventional reformers).

To mention, Barbiery et al. performed SMR reaction using a membrane reactor which generates pure hydrogen [111]. The membrane reactor comprised of two tubular membranes: one Pd based and another made from porous alumina. The reactor run at 350–500 °C with no sweep gas, and the steam/methane molar ratio was in the range 3.5–5.9. A 7 % increase in methane conversion over the thermodynamic equilibrium value was observed for the membrane.

Membrane performance is adversely affected by the presence of other gases and thermal cycling. For instance, the gases such as N_2, steam, CO, and CO_2 can significantly decrease the permeability of the membrane. Such negative effect on the permeation rate is attributed to the competitive adsorption of these gases with the hydrogen [112]. However, at higher temperature (>500 °C), such effects can be diminished [113]. Therefore, at normal working temperatures (500–600 °C), these inhibition effects will not be significant.

Membrane reactors are able to achieve higher methane conversion at lower temperature and higher total efficiency compared to the conventional SMR technique. However, more research needs to be conducted to enable the production of a membrane that combines a thin layer, high flux, and stability in all operating conditions. Membrane materials other than Pd could provide an inexpensive alternative and should be explored.

3.4 In Situ CO_2 Capture Using NaOH

As mentioned briefly in Sect. 2.2.1 of Chap. 2, sodium hydroxide when included in the conventional SMR technique can simultaneously capture CO_2 and generate hydrogen. Such modification may lead to less energy or carbon penalty. The inclusion of sodium hydroxide produces not only hydrogen as a product, but also soda ash, which has huge commercial value. A similar approach called Skymine®

technology (Skyonic) has received a lot of interest and is currently under deployment. However, such method has its own inherent disadvantages and will be discussed in the later part of this section.

We also performed a series of work at laboratory scale to illustrate the feasibility of modified steam methane reformation (MSMR) (CH$_4$ + NaOH + H$_2$O = Na$_2$CO$_3$ + H$_2$) reaction [114]. Moreover, a detailed catalytic study was also performed and can be found elsewhere [115]. The result showed that the MSMR is a single-step reaction and can yield 98 % conversion at 600 °C under 25 ml/min of CH$_4$. The catalytic effect on the conversion of NaOH to Na$_2$CO$_3$ is most pronounced at low temperatures (300 °C), but the effect decreases as temperature increases to 600 °C. With the use of variously milled nickel catalysts, the reaction temperature can be further lowered. Using Scherrer equation, the crystallite size of variously milled nickel catalysts was calculated. The crystallite sizes of raw, 2-h-, 3-h-, and 4-h-milled nickel particles were 304.8, 265.7, 239.9, and 184.7 Å, respectively, whereas the average particle size increased from 3–4 μm (raw nickel) to 33–38 μm (4-h-milled nickel). It is obvious that prolonged milling of nickel catalysts (time = 4 h) caused the particles to coalesce and grow in size decreasing the reaction rate. The best catalytic performance was exhibited by 2-h-milled nickel catalysts. The study showed that then decrease in catalytic activity of nickel is dependent on the crystallite size and thus milling time. Here, a detailed explanation is provided to clarify the phenomenon behind the reduction in catalytic performance of nickel upon prolonged milling.

Figure 3.7 shows the schematic of the morphology transformation observed for nickel catalysts at different milling time.

It is well known that the mechanical milling is controlled by the two processes—(i) cold welding and (ii) fracture. The particle size and morphology of catalysts are determined by the competition between cold welding and fracture process. As can be seen here, the milling process is dominated by the cold welding process. On milling, nickel powders are cold welded and eventually form elongated powders. The formation of disk-shaped particles demonstrates the severe ball–powder interaction induces large stress for the nickel powders. Considering that the morphology and the thickness of the disk-shaped nickel catalysts are quite similar to those of sticked powder, it seems that the disk-shaped particles are formed by breaking off of the powder sticked on the ball surfaces and the mill container walls (due to spalling action of the ball). Flake-shaped particles have high diameter-to-thickness aspect ratio. Due to its high aspect ratio, a flake particle has a larger specific surface area than a spherical particle of the same volume, which can increase the chemical reactivity [116, 117]. The heat inside the milling jar increases with increase in milling time. Hence, the agglomeration occurs causing the particle size to increase accordingly. Here, the particle size for 3-h or 4-h-milled samples is relatively larger than 0-h or 2-h-milled particles. As a consequence, the aspect ratio of nickel particles drastically decreases for more than 2 h milling time. Similar agglomeration behavior has been previously reported for ductile metals such as Cu and Ni [118, 119].

We also explored the behavior of different alkali for MSMR technique [120]. Figure 3.8 shows the effect of alkali hydroxide on the methane transformation rate at 400 °C for three alkali hydroxides.

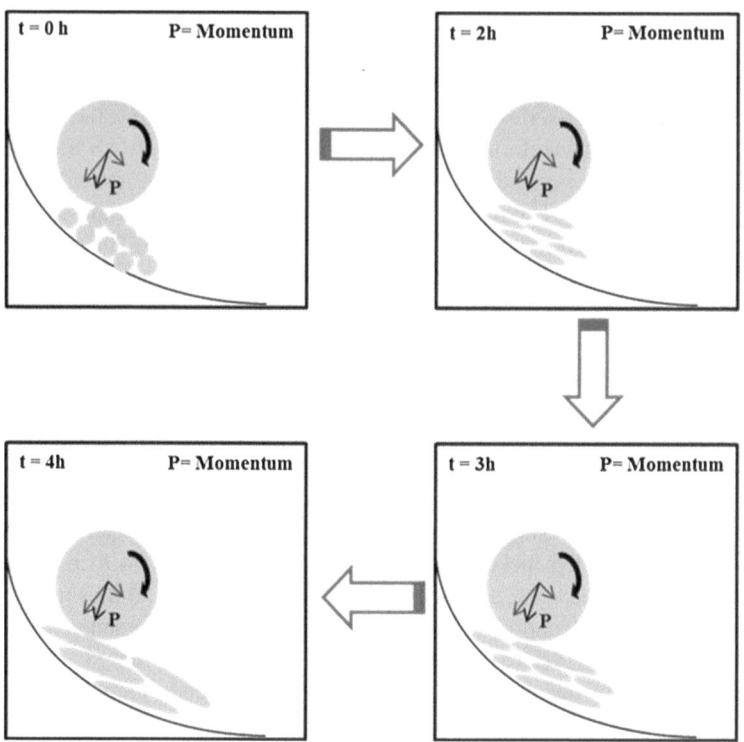

Fig. 3.7 Schematics of the transformation of nickel particle shape depending on the milling time
(*t* ball-milling time)

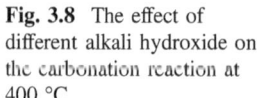

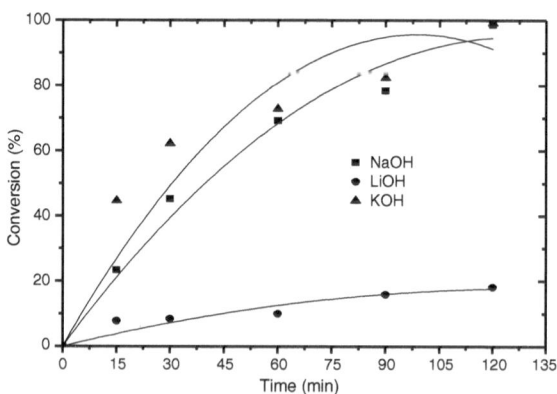

Fig. 3.8 The effect of different alkali hydroxide on the carbonation reaction at 400 °C

The carbonate formation rate becomes higher in the following order: KOH ≈ NaOH > LiOH. The similar observation of hydroxide reactivity is reported by Ishida et al. [121] for the coal–alkali hydroxide–steam reaction and by Kamo et al. [122] for PVC or activated carbon steam gasification reaction. Ishida et al.

[121] correlated such a change in hydroxide reactivity with their melting points. In this case, the reaction occurs with high rate well below the melting point of the hydroxides. Thus, change in reactivity of alkali hydroxides can be attributed to their alkalinity rather than to melting point differences.

Also, it is noteworthy to check the economic viability of MSMR process. For instance, when 80 tonnes of NaOH reacts with 16 tonnes of methane, we can achieve maximum of 8 tonnes H_2 and 106 tonnes of soda. If the efficiency of the process, capital, operational, and other costs is not taken into account, then the price difference between soda and the hydroxide would largely determine the production cost of hydrogen [123]. One such method for calculation is shown in Fig. 3.9. It suggests that if there is not much difference between the price of the product soda and the reactant hydroxide, a plant owner with access to the hydroxide can gain substantial benefit. If the demand for soda and hydrogen decreases, then the MSMR technique would not be profitable. However, if the proposed method is employed, it seems certain that the price of soda increases due to the demand of hydrogen.

As mentioned in Chap. 2, sodium hydroxide is produced by the electrolysis of brine as [as shown in (3.10)]

$$2NaCl + 2H_2O \rightarrow Cl_2 + H_2 + 2NaOH \qquad (3.10)$$

The reaction generates almost equal amount of chlorine and sodium hydroxide. The degree of demand would determine which compound (chlorine or NaOH) can be regarded as a by-product, and the price will vary accordingly. The price fluctuations can be extreme: In times of oversupply, caustic soda prices can be as low as $20–30 per tonne, whereas in short supply, prices can be $300 or higher per tonne [124]. It would be wise to convert the caustic soda to soda in order to offset the vagary of the price fluctuation as caustic soda can be replaced by soda ash in many

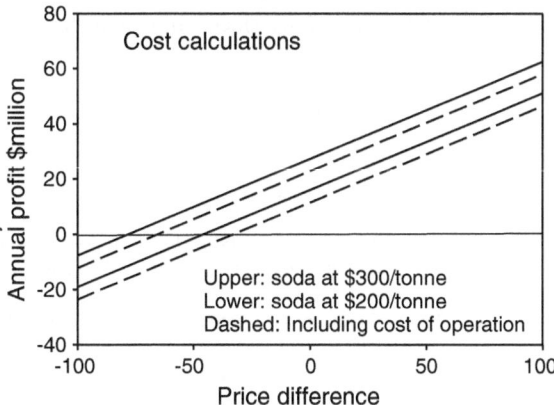

Fig. 3.9 The price difference in soda and sodium hydroxide determines the price of hydrogen. In this example, the price of hydrogen is fixed at $1,000/tonne. If the price of soda goes below the sodium hydroxide by $30/tonne, the method may not be profitable

applications, especially in the pulp and paper, water treatment, and certain chemical sectors [125].

It is unfortunate that the chlor-alkali plants produce more carbon dioxide than is demanded. The chemistry of the electrolysis process is bound to emit huge carbon dioxide. Moreover, the number of chlor-alkali plants may even grow all over the developing countries. The above-mentioned method reduces the emission of these plants and can also provide benefit to the plant owners. In addition, the use of clean hydrogen will benefit the environment.

In summary, there is a way to produce hydrogen with zero emission by using the sodium hydroxide from existing chlor-alkali plants. This modified SMR process removes a major impediment in the use of hydrogen for vehicular transportation in a commercially viable way.

3.5 Conclusion

It is encouraging to see that currently various novel methods are being investigated and developed to reduce the carbon emission emitted by the SMR technique. These methods are primarily based on the use of metal oxides or alkali as CO_2 sorbent, renewable energy sources, and membranes. In this chapter, the shortcomings for these novel methods are presented with their proposed solutions. Unfortunately, for most of these methods, fossil fuels continue to be a prime energy source. Overall, there are some challenges to overcome before these proposed methods can be deployed at a large scale. Owing to the scale of global energy demand, renewable energy-based methods can be stationed in the near-to-far term. However, it is important to concurrently develop all these methods as there will not be any single solution. Rather, optimal solution may vary with regions depending on their energy source, availability of raw materials, and end use of hydrogen. It is very critical to establish net reduction in CO_2 emissions since only then the continuous use of hydrogen can be justified.

References

1. Adris AM, Pruden BB, Lim CJ, Grace JR (1996) On the reported attempts to radically improve the performance of steam methane reforming reactor. Candian J Chem Eng 74:177–186. doi:10.1002/cjce.5450740202
2. Goetsch DA, Say GR (1989) Synthesis gas preparation and catalyst thereof. US Patent 4,877,550
3. Minet RG, Tsotsis TT (1991) Catalytic ceramic membrane steam hydrocarbon reformer. US Patent 4,981,676
4. Paloumbis S, Petersen EE (1982) Coke deposition on a commercial nickel oxide catalyst during the steam reforming of methane. Chem React Eng ACS Symp Ser 38:489–494. doi:10.1021/bk-1982-0196.ch038

5. Singh CPP, Saraf DN (1979) Simulation of side-fired steam hydrocarbon reformers. Ind Eng Chem Process Des Dev 18:1–7. doi:10.1021/i260069a001
6. Hyman M (1968) Simulate methane reformer reactions. Hydrog Proc 47:131–137
7. Reichel W, Lippert H (1984) Wirbelbett-Reaktorsystem. Deutsches Patentamt No DE. 3331202
8. Spagnolo DA, Cornett LJ, Chuang KT (1992) Direct electro-steam reforming: a novel catalytic approach. Int J Hydrog Energy 17:839–846. doi:10.1016/0360-3199(92)90033-S
9. Goetsch DA, Say GR, Vargas JM, Eberly PE (1989) Synthesis gas preparation and catalyst therefore. US Patent 4,888,131
10. Guerrieri SA (1970) Steam reforming of hydrocarbons. US Patent 3,524,819
11. Robinson LF (1980) Reforming of hydrocarbons. US Patent 4,224,298
12. Suzumura H, Makihara H (1986) Manufacture of hydrogen-containing gas. Japanese Patent JP61186201
13. Adris AM, Pruden BB, Lim CJ, Grace JR (1996) On the reported attempts to radically improve the performance of the steam methane reforming reactor. Can J Chem Eng 74:177–186. doi:10.1002/cjce.5450740202
14. Brun-Tsekhovoi AR, Zadorin AR, Katsobashvili YR, Kourdumov SS (1988) The process of catalytic steam-reforming of hydrocarbons in the presence of carbon dioxide acceptor. In: Hydrogen energy progress VII, proceedings of the 7th world hydrogen energy conference, pp. 885–900
15. Gupta RB (2009) Hydrogen fuel: production, transport, and storage. CRC Press, Boca Raton, p. 9 (Chapter 1)
16. LaBar MP (2002) The gas-turbine modular helium reactor: a promising option for near-term deployment. In: Proceedings of international congress on advanced nuclear power plants. Florida
17. Muradov N (2009) Production of hydrogen from hydrocarbons. In: Gupta R (ed) Hydrogen fuel, production, transport and storage. Boca Raton
18. Bolat P, Thiel C (2014) Hydrogen supply chain architecture for bottom-up energy systems models. Part 2: techno-economic inputs for hydrogen production pathways. Int J Hydrog Energy 39:8898–8925. doi:10.1016/i.ijhydene.2014.03.170
19. Farbman GH (1979) Hydrogen production by the westinghouse sulfur cycle process: program status. Int J Hydrog Energy 4:111–122. doi:10.1016/0360-3199(79)90045-4
20. Yildiz B, Kazimi MS (2006) Efficiency of hydrogen production systems using alternative nuclear energy technologies. Int J Hydrog Energy 31:77–92. doi:10.1016/i.ijhydene.2005.02.009
21. Ohashi H, Inaba Y, Nishihara T, Takeda T, Hayashi K, Takada S, Inagaki Y (2006) Development of control technology for HTTR hydrogen production system with mock-up test facility—system controllability test for loss of chemical reaction. Nucl Eng Des 236:1396–1410. doi:10.1016/j.nucengdes.2006.01.005
22. Steinfeld A, Meier A (2004) Solar fuels and materials. Encycl Energy 5:623–637 (Cleveland (ed))
23. Edwards JH, Do KT, Maitra AM, Schuck S, Fok W, Stein W (1996) The use of solar-based CO_2/CH_4 reforming for reducing greenhouse gas emissions during the generation of electricity and process heat. Energy Convers Manage 37:1339–1344. doi:10.1016/0196-8904(95)00343-6
24. Jensen J, Poulsen J, Andersen N (2010) From coal to clean energy. Nitrogen + Syngas 310
25. Aristov YI, Fedoseev VI, Parmon VN (1997) High-density conversion of light energy via direct illumination of catalyst. Int J Hydrog Energy 22:869–874. doi:10.1016/S0360-3199(96)00238-8
26. Yokota O, Oku Y, Arakawa M, Hasegawa N, Matsunami J, Kaneko H, Tamaura Y, Kitamura M (2000) Steam reforming of methane using a solar simulator controlled by H2O/CH4 = 1/1. Appl Organomet Chem 14:867–870. doi:10.1002/1099-0739(200012)14:12<867:AID-AOC99>3.0.CO;2-X

27. Lau C, Tsolakis A, Wyszynsk ML (2011) Biogas upgrade to syn-gas (H_2–CO) via dry and oxidative reforming. Int J Hydrog Energy 36:397–404. doi:10.1016/j.ijhydene.2010.09.086

28. Rasi S, Veijanen A, Rintala J (2007) Trace compounds of biogas from different biogas production plants. Energy 32:1375–1380. doi:10.1016/j.energy.2006.10.018

29. Xie M, Zhou Z, Qi Y, Cheng Z, Yuan W (2012) Sorption-enhanced steam methane reforming by in situ CO_2 capture on a CaO–$Ca_9Al_6O_{18}$ sorbent. Chem Eng J 207–208:142–150. doi:10.1016/j.cej.2012.06.032

30. Ding Y (2000) Adsorption-enhanced steam-methane reforming. Chem Eng Sci 55:3929–3940. doi:10.1016/S0009-2509(99)00597-7

31. Jakobsen HA 2008 Chemical reactor modeling. Multiphase reactive flows. Springer, Berlin. pp 659–677

32. Balasubramanian B, Ortiz LA, Kaytakoglu S, Harrison DP (1999) Hydrogen from methane in a single-step process. Chem Eng Sci 54:3543–3552. doi:10.1016/S0009-2509(98)00425-4

33. Wang Y, Chao Z, Chen D, Jakobsen HA (2011) SE-SMR process performance in CFB reactors: simulation of the CO_2 adsorption/desorption processes with CaO based sorbents. Int J Greenhouse Gas Control 5:489–497. doi:10.1016/i.ijggc.2010.09.001

34. Solsvik J, Jakobsen HA (2011) A numerical study of a two property catalyst/sorbent pellet design for the sorption-enhanced steam–methane reforming process: modeling complexity and parameter sensitivity study. Chem Eng J 178:407–422. doi:10.1016/j.cej.2011.10.045

35. Lindborg H, Jakobsen HA (2009) Sorption enhanced steam methane reforming process performance and bubbling fluidized bed reactor design analysis by use of a two-fluid model. Ind Eng Chem Res 48:1332–1342. doi:10.1021/ie800522p

36. Wang J, Wang Y, Jakobsen HA (2014) The modeling of circulating fluidized bed reactors for SE-SMR process and sorbent regeneration. Chem Eng Sci 108:57–65. doi:10.1016/j.ces.2013.12.012

37. Yi BK, Harrison DP (2005) Low-pressure sorption-enhanced hydrogen production. Ind Eng Chem Res 44:1665–1669. doi:10.1021/ie048883g

38. Johnsen K, Ryu HJ, Grace JR, Lim CJ (2006) Sorption-enhanced steam reforming of methane in a fluidized bed reactor with dolomite as CO_2-acceptor. Chem Eng Sci 61:1195–1202. doi:10.1016/j.ces.2005.08.022

39. Silaban A, Narcida M, Harrison DP (1996) Characteristics of the reversible reaction between CO_2 (g) and calcined dolomite. Chem Eng Comm 146:149–162. doi:10.1080/00986449608936487

40. Ochoa-Fernández E, Rusten HK, Jakobsen HA, Ronning M, Holmen A, Chen D (2005) Catal Today 106:41–46. doi:10.1016/j.cattod.2005.07.146

41. Zhao T, Ochoa-Fernández E, Ronning M, Chen D (2007) Preparation and high-temperature CO_2 capture properties of nanocrystalline Na_2ZrO_3. Chem Mater 19:3294–3301. doi:10.1021/cm062732h

42. Essaki K, Muramatsu T, Kato M (2008) Effect of equilibrium shift by using lithium silicate pellets in methane steam reforming. Int J Hydrog Energy 33:4555–4559. doi:10.1016/j.ijhydene.2008.05.063

43. Ding Y, Alpay E (2000) Adsorption-enhanced steam-methane reforming. Chem Eng Sci 55:3929–3940. doi:10.1016/S0009-2509(99)00597-7

44. Halabi MH, de Croon MHJM, Schaff JVD, Cobden PD, Schouten JC (2012) A novel catalyst–sorbent system for an efficient H_2 production with in-situ CO_2 capture. Int J Hydrog Energy 37:4987–4996. doi:10.016/j.ijhydene.2011.12.025

45. Kumar S (2014) A comparative study of CO_2 sorption properties for different oxides. Mater Renew Sus Energy 3:30. doi:10.1007/s40243-014-0030-9

46. Ochoa-Fernández E, Haugen G, Zhao T, Ronning M, Aartun I, Borresen B, Rytter E, Ronnekleiv M, Chen D (2007) Process design simulation of H_2 production by sorption enhanced steam methane reforming: evaluation of potential CO_2 acceptors. Green Chem 9:654–662. doi:10.1039/B614270B

47. Broda M, Manovic V, Imtiaz Q, Kierzkowska AM, Anthony EJ, Muller CR (2013) Reforming reaction over a synthetic CaO-based sorbent and a Ni catalyst. Environ Sci Technol 47:6007–6014. doi:10.1021/es305113p
48. Harrison DP (2008) Sorption-enhanced hydrogen production: a review. Ind Eng Chem Res 47:6486–6501. doi:10.1021/ie800298z
49. Anthony EJ (2011) Ca looping technology: current status, developments and future directions. Greenhouse Gas Sci Technol 1:36–47. doi:10.1002/ghg3.2
50. Judd MD, Pope MI (1970) Formation and surface properties of electron-emissive coatings V. DTA, ETA and dilatometry studies on some alkaline–earth carbonates. Appl Chem 20:384–388. doi:10.1002/jctb.5010201205
51. Manovic V, Anthony EJ (2008) Thermal activation of CaO-based sorbent and self-reactivation during CO_2 capture looping cycles. Environ Sci Technol 42:4170–4174. doi:10.1021/es800152s
52. Manovic V, Anthony EJ (2007) Steam reactivation of spent CaO-based sorbent for multiple CO_2 capture cycles. Environ Sci Technol 41:1420–1425. doi:10.1021/es0621344
53. Materić V, Sheppard C, Smedley SI (2010) Effect of repeated steam hydration reactivation on CaO-based sorbents for CO_2 capture. Environ Sci Technol 44:9496–9501. doi:10.1021/es102623k
54. Kierzkowska AM, Muller CR (2013) Sol–gel-derived, calcium-based, copper-functionalised CO_2 sorbents for an integrated chemical looping combustion–calcium looping CO_2 capture process. ChemPlusChem 78: 92−100. doi:10.1002/cplu.201200232
55. Pacciani R, Muller CR, Davidson JF, Dennis JS, Hayhurst AN (2008) How does the concentration of CO_2 affect its uptake by a synthetic Ca-based solid sorbent? AIChE J 54:3308–3311. doi:10.1002/aic.11611
56. Broda M, Kierzkowska AM, Muller CR (2012) Influence of the calcination and carbonation conditions on the CO_2 uptake of synthetic Ca-based CO_2 sorbents. Environ Sci Technol 46:10849–10856. doi:10.1021/es302757e
57. Broda M, Muller CR (2012) Synthesis of highly efficient, Ca-based, Al_2O_3-stabilized, carbon gel-templated CO_2 sorbents. Adv Mater 24:3059–3064. doi:10.1002/adma.201104787
58. Radfarnia HR, Iliuta MC (2012) Development of zirconium-stabilized calcium oxide absorbent for cyclic high-temperature CO_2 capture. Ind Eng Chem Res 51:10390–10398. doi:10.1021/ie301287k
59. Filitz R, Kierzkowska AM, Broda M, Muller CR (2012) Highly efficient CO_2 sorbents: development of synthetic, calcium-rich dolomites. Environ Sci Technol 46:559–565. doi:10.1021/es2034697
60. Sultan DS, Muller CR, Dennis JS (2010) Capture of CO_2 using sorbents of calcium magnesium acetate (CMA). Energy Fuels 24:3687–3697. doi:10.1021/ef100072q
61. Chanburanasiri N, Ribeiro AM, Rodrigues AE, Arpornwichanop A, Laosiripojana N, Praserthdam P, Assabumrungrat S (2011) Hydrogen production via sorption enhanced steam methane reforming process using Ni/CaO multifunctional catalyst. Ind Eng Chem Res 50:13662–13671. doi:10.1021/ie201226j
62. Li Y, Zhao C, Chen H, Duan L, Chen X (2010) Cyclic CO_2 capture behavior of $KMnO_4$-doped CaO-based sorbent. Fuel 89:642–649. doi:10.1016/j.fuel.2009.08.041
63. Wu SF, Zhu YQ (2010) Behavior of $CaTiO_3$/Nano-CaO as a CO_2 reactive adsorbent. Ind Eng Chem Res 49:2701–2706. doi:10.1021/ie900900r
64. Li ZS, Cai N, Huang Y, Han H (2005) Synthesis, experimental studies, and analysis of a new calcium-based carbon dioxide absorbent. Energy Fuels 19:1447–1452. doi:10.1021/ef0496799
65. Martzavaltzi CS, Lemonidou AA (2008) Development of new CaO based sorbent materials for CO_2 removal at high temperature. Microporous Mesoporous Mater 110:119–127. doi:10.1016/j.micromeso.2007.10.006
66. Barelli L, Bidini G, Michele A, Gallorini F, Petrillo C, Sacchetti F (2014) Synthesis and test of sorbents based on calcium aluminates for SE-SR. Appl Energy 127:81–92. doi:10.1016/j.apenergy.2014.04.034

67. Choi S, Drese JH, Jones CW (2009) Adsorbent materials for carbon dioxide capture from large anthropogenic point sources. ChemSusChem 2:796–854. doi:10.1002/cssc.200900036
68. Wang Q, Luo J, Zhong Z, Borqna A (2011) CO_2 capture by solid adsorbents and their applications: current status and new trends. Energy Environ Sci 4:42–55. doi:10.1039/C0EE00064G
69. Drage TC, Snape CE, Stevens LA, Wood J, Wang J, Cooper AI (2012) Materials challenges for the development of solid sorbents for post-combustion carbon capture. J Mater Chem 22:2815–2823. doi:10.1039/C2JM12592G
70. Ebner AD, Reynolds SP, Ritter JA (2007) Nonequilibrium kinetic model that describes the reversible adsorption and desorption behavior of CO_2 in a K-promoted hydrotalcite-like compound. Ind Eng Chem Res 46:1737–1744. doi:10.1021/ie061042k
71. Reddy MK, Xu ZP, Lu M, Diniz da Costa JC (2008) Influence of water on high-temperature CO_2 capture using layered double hydroxide derivatives. Ind Eng Chem Res 47:2630–2635. doi:10.1021/ie0716060
72. Ishida M, Jin H (1994) A new advanced power-generation system using chemical-looping combustion. Energy 19:415–422. doi:10.1016/0360-5442(94)90120-1
73. Anheden A, Svedberg G (1998) Exergy analysis of chemical-looping combustion systems. Energy Convers Manage 39:1967–1980. doi:10.1016/S0196-8904(98)00052-1
74. Gnanapragasam NV, Reddy BV, Rosen MA (2009) Hydrogen production from coal using coal direct chemical looping and syngas chemical looping combustion systems: assessment of system operation and resource requirements. Int J Hydrog Energy 34:2606–2615. doi:10.1016/j.ijhydene.2009.01.036
75. Ramkumar S, Fan LS (2010) Calcium looping process (CLP) for enhanced noncatalytic hydrogen production with integrated carbon dioxide capture. Energy Fuel 24:4408–4418. doi:10.1021/ef100346j
76. Corbella BM, de Diego LF, Garcia-Labiano F, Adanez J, Palacios JM (2006) Performance in a fixed-bed reactor of titania-supported nickel oxide as oxygen carriers for the chemical-looping combustion of methane in multicycle tests. Ind Eng Chem Res 45:157–165. doi:10.1021/ie050756c
77. Murugan A, Thursfield A, Metcalfe AS (2011) A chemical looping process for hydrogen production using iron-containing perovskites. Energy Env Sci 4:4639–4649. doi:10.1039/C1EE02142G
78. Dueso C, Garca-Labiano F, Adnez J, de Diego L, Gayn P, Abad A (2009) Syngas combustion in a chemical-looping combustion system using an impregnated Ni-based oxygen carrier. Fuel 88:2357–2364. doi:10.1016/j.fuel.2008.11.026
79. Kolbitsch P, Pröll T, Bolhar-Nordenkampf J, Hofbauer H (2009) Design of a chemical looping combustor using a dual circulating fluidized bed (DCFB) reactor system. Chem Eng Technol 32:398–403. doi:10.1002/ceat.200800378
80. Ishida M, Jin H, Okamoto T (1998) Kinetic behavior of solid particle in chemical-looping combustion: suppressing carbon deposition in reduction. Energy Fuel 12:223–229. doi:10.1021/ef970041p
81. Erri P, Varma A (2007) Spinel-supported oxygen carriers for inherent CO_2 separation during power generation. Ind Eng Chem Res 46:8597–8601. doi:10.1021/ie070068o
82. Ryden M, Lyngfelt A (2006) Using steam reforming to produce hydrogen with carbon dioxide capture by chemical-looping combustion. Int J Hydrog Energy 31:1271–1283. doi:10.1016/j.ijhydene.2005.12.003
83. Gayan P, Dueso C, Abad A, Adanez J, Diego L, Garcia-Labiano F (2009) NiO/Al_2O_3 oxygen carriers for chemical-looping combustion prepared by impregnation and deposition–precipitation methods. Fuel 88:1016–1023. doi:10.1016/j.fuel.2008.12.007
84. Corbella BM, de Diego LF, García-Labiano F, Adánez J, Palacios JM (2005) Characterization study and five-cycle tests in a fixed-bed reactor of titania-supported nickel oxide as oxygen carriers for the chemical-looping combustion of methane. Environ Sci Technol 39:5796–5803. doi:10.1021/es048015a

85. Saha C, Roy B, Bhattacharya S (2011) Chemical looping combustion of Victorian brown coal using NiO oxygen carrier. Int J Hydrog Energy 36:3253–3259. doi:10.1016/j.ijhydene. 2010.11.119

86. Mattisson T, Johansson M, Lyngfelt A (2006) The use of NiO as an oxygen carrier in chemical-looping combustion. Fuel 85:736–747. doi:10.1016/j.fuel.2005.07.021

87. Mattisson T, Lyngfelt A, Cho P (2001) The use of iron oxide as an oxygen carrier in chemical-looping combustion of methane with inherent separation of CO_2. Fuel 80:1953–1962. doi:10. 1016/S0016-2361(01)00051-5

88. Abad A, Mattisson T, Lyngfelt A, Johansson M (2007) The use of iron oxide as oxygen carrier in a chemical-looping reactor. Fuel 86:1021–1035. doi:10.1016/j.fuel.2006.09.021

89. Cho P, Mattisson T, Lyngfelt A (2005) Carbon formation on nickel and iron oxide-containing oxygen carriers for chemical-looping combustion. Ind Eng Chem Res 44:668–676. doi:10.1021/ie049420d

90. Wang B, Gao C, Wang W, Zheng C (2011) Chemical looping combustion of coal with CuO-Fe mechanically mixed oxygen carrier. Proc Eng 16:48–53. doi:10.1016/j.proeng.2011.08. 1050

91. Ortiz M, Gayán P, de Diego LF, García-Labiano F, Abad A, Pans MA, Adánez J (2011) Hydrogen production with CO_2 capture by coupling steam reforming of methane and chemical-looping combustion: Use of an iron-based waste product as oxygen carrier burning a PSA tail gas. J Power sources 196:4370–4381. doi:10.1016/j.jpowsour.2010.09.101

92. Cho W, Seo M, Kim S, Kang K, Bae K, Kim C, Jeong S, Park C (2012) Reactivity of iron oxide as an oxygen carrier for chemical-looping hydrogen production. Int J Hydrog Energy 37:16852–16863. doi:10.1016/j.ijhydene.2012.08.020

93. Zhu X, Wei Y, Wang H, Li K (2013) Ce–Fe oxygen carriers for chemical-looping steam methane reforming. Int J Hydrog Energy 38:4492–4501. doi:10.1016/j.ijhydene.2013.01.115

94. Corbella Beatríz M, Palacios José María (2007) Titania-supported iron oxide as oxygen carrier for chemical-looping combustion of methane. Fuel 86:113–122. doi:10.1016/j.fuel. 2006.05.026

95. Wang C, Zhao H, Zheng Y, Liu Z, Yan R, Zheng C (2012) Chemical looping combustion of a Chinese anthracite with Fe_2O_3-based and CuO-based oxygen carriers. Fuel Process Technol 96:104–115. doi:10.1016/j.fuproc.2011.12.030

96. de Diego LF, García-Labiano F, Adánez J, Gayán P, Abad A, Corbella BM, Palacios JM (2004) Development of Cu-based oxygen carriers for chemical-looping combustion. Fuel 83:1749–1757. doi:10.1016/j.fuel.2004.03.003

97. Saha C, Bhattacharya S (2011) Comparison of CuO and NiO as oxygen carrier in chemical looping combustion of a Victorian brown coal. Int J Hydrog Energy 36:12048–12057. doi:10.1016/j.ijhydene.2011.06.065

98. Zheng X, Su Q, Mi W, Zhang P (2014) Effect of steam reforming on methane-fueled chemical looping combustion with Cu-based oxygen carrier. Int J Hydrog Energy 39:9158–9168. doi:10.1016/j.ijhydene.2014.03.245

99. Abad A, Mattisson T, Lyngfelt A, Rydén M (2006) Chemical-looping combustion in a 300 W continuously operating reactor system using a manganese-based oxygen carrier. Fuel 85:1174–1185. doi:10.1016/j.fuel.2005.11.014

100. Zafar Q, Abad A, Mattisson T, Gevert B, Strand M (2007) Reduction and oxidation kinetics of Mn_3O_4/Mg–ZrO_2 oxygen carrier particles for chemical-looping combustion. Chem Eng Sci 62:6556–6567. doi:10.1016/j.ces.2007.07.011

101. Hossain MM, Sedor KE, de Lasa HI (2007) Co–Ni/Al_2O_3 oxygen carrier for fluidized bed chemical-looping combustion: desorption kinetics and metal–support interaction. Chem Eng Sci 62:5464–5472. doi:10.1016/j.ces.2006.12.066

102. Tian H, Guo Q, Chang J (2008) Investigation into decomposition behavior of $CaSO_4$ in chemical-looping combustion. Energy Fuel 22:3915–3921. doi:10.1021/ef800508w

103. Song Q, Xiao R, Deng Z, Zheng W, Shen L, Xiao J (2008) Multicycle Study on chemical-looping combustion of simulated coal gas with a $CaSO_4$ oxygen carrier in a fluidized bed reactor. Energy Fuel 22:3661–3672. doi:10.1021/ef800275a

104. Bhavsar S, Tackett B, Veser G (2014) Evaluation of iron- and manganese-based mono- and mixed-metallic oxygen carriers for chemical looping combustions. Fuel 136:268–279. doi:10.1016/j.fuel.2014.07.068

105. Zhao K, He F, Huang Z, Zheng A, Li H, Zhao Z (2014) Three-dimensionally ordered macroporous LaFeO₃ perovskites for chemical-looping steam reforming of methane. Int J Hydrog Energy 39:3243–3252. doi:10.1016/j.ijhydene.2013.12.046

106. Murugan A (2011) Iron-containing perovskite materials for stable hydrogen production by chemical looping water splitting. New Castle University, UK

107. Roses L, Gallucci F, Manzolini G, Annaland MS (2013) Experimental study of steam methane reforming in a Pd-based fluidized bed membrane reactor. Chem Eng J 222:307–320. doi:10.1016/j.cej.2013.02.069

108. Xie D, Yu J, Wang F, Zhang N, Wang W, Yu H, Peng F, Park AA (2011) Hydrogen permeability of Pd–Ag membrane modules with porous stainless steel substrates. Int J Hydrog Energy 36:1014–1026. doi:10.1016/j.ijhydene.2010.10.030

109. Chen WH, Syu WZ, Hung CI, Lin YL, Yang CC (2012) A numerical approach of conjugate hydrogen permeation and polarization in a Pd membrane tube. Int J Hydrog Energy 37:12666–12679. doi:10.1016/j.ijhydene.2012.05.128

110. Ledjeff-Hey K, Formanski V, Kalk Th, Roes J (1998) Compact hydrogen production systems for solid polymer fuel cells. J Power Sources 71:199–207. doi:10.1016/S0378-7753(97) 02760-2

111. Damle S (2001) Recovery of carbon dioxide in advanced fossil fuel conversion processes using a membrane reactor. In: First National Conference on Carbon Sequestration. Washington

112. Hou K, Hughes R (2002) The effect of external mass transfer, competitive adsorption and coking on hydrogen permeation through thin Pd/Ag membranes. J Membr Sci 206:119–130. doi:10.1016/S0376-7388(01)00770-0

113. Bus E (2002) Poisoning of Palladium membranes during separation of hydrogen from CPO-WGS product streams. Utrecht University

114. Saxena S, Kumar S, Drozd V (2011) A modified steam-methane-reformation reaction for hydrogen production. Int J Hydrog Energy 36:4366–4369. doi:10.1016/j.ijhydene.2010.12.133

115. Kumar S, Drozd V, Saxena S (2012) A modified method for production of hydrogen from methane. Int J Energy Res 36:1133–1138. doi:10.1002/er.1854

116. Trunov MA, Schoenitz M, Zhu X, Dreizin EL (2005) Effect of polymorphic phase transformations in Al₂O₃ film on oxidation kinetics of aluminum powders. Combust Flame 140:310–318. doi:10.1016/j.combustflame.2004.10.010

117. Cashdollar KL (2000) Overview of dust explosibility characteristics. J Loss Prev Process Ind 13:183–199. doi:10.1016/S0950-4230 (99)00039-X

118. Lee GG, Hashimoto H, Watanabe R (1995) Development of particle morphology during dry ball milling of Cu powder. Mater Trans 36:548–554

119. Cho DG, Yang SK, Lee JS, Lee CS (2011) Investigation of mechanical properties and elongated Ni grain growth in an Al₂O₃-Ni composite during low-energy ball milling. Mater Trans 52:2131–2136

120. Kumar S (2013) Clean hydrogen production and carbon dioxide capture methods. FIU Electronic Theses and Dissertations. Paper 1039 http://digitalcommons.fiu.edu/etd/1039

121. Ishida M, Toida M, Shimizu T, Takenaka S, Otsuka K (2004) Formation of hydrogen without COₓ from carbon, water, and alkali hydroxide. Ind Eng Chem Res 43:7204–7206. doi:10.1021/ie049360b

122. Kamo T, Takaoka K, Otomo J, Takahashi H (2006) J Mater Cycles Waste Manage 8:109–115. doi:10.1007/s10163-006-0152-y

123. Eurochlor report 1997. http://www.eurochlor.org/

124. http://www.icis.com/Articles/2009/12/30/9321358/OUTLOOK-10-US-chlor-alkali-on-a-tightrope.html

125. Dennis S, Kostick D (1998) Soda Ash US Geological Survey, Mineral Commodity Summaries. http://minerals.usgs.gov/minerals/pubs/commodity/soda_ash/610398.pdf

Chapter 4
Modified Coal Gasification Process for Hydrogen Production

Abstract A significant amount of work has been performed to modify the conventional coal gasification process. This chapter presents a brief review of the methods that are supposed to reduce the vast emission of carbon dioxide during the process. Most of these proposed methods include CaO as a CO_2 sorbent. Moreover, efforts have also been made to integrate gasification of coal and biomass which can offer several advantages. Similar to Chap. 3, the present chapter briefs about the addition of sodium hydroxide (as a reactant) to the conventional coal gasification process. The sodium hydroxide-assisted reaction operates at a relatively mild condition and has potential to substitute the conventional method. However, there are several existing issues related to the proposed technique which needs to be resolved prior to its deployment.

Keywords Coal gasification · Sorbent · Biomass

4.1 Introduction

As coal reserves are available only for a definite time period, it must be used effectively. Coal gasification is a chemical process in which solid coal reacts with high pressure and high temperature steam and oxygen to form a synthetic gaseous mixture of hydrocarbons, which can be used as a gaseous fuel or can be refined to produce hydrogen gas. However, the hydrogen production using steam coal gasification process suffers severe disadvantages as mentioned in Sect. 2.1.1. Thus, new technologies have to be developed to improve the efficiency of hydrogen production and capturing carbon dioxide during hydrogen production. In the following section, innovative techniques are listed which intend to produce clean hydrogen using coal.

© The Author(s) 2015
S. Kumar, *Clean Hydrogen Production Methods*,
SpringerBriefs in Energy, DOI 10.1007/978-3-319-14087-2_4

4.2 Coal Gasification Using Fossil Energy Sources

4.2.1 HyPr-RING

It is well known that the conventional steam coal gasification $(C + H_2O \rightarrow CO + H_2)$ is an endothermic process and requires high temperature (>1,000 °C), whereas water–gas shift reaction $(CO + H_2O \rightarrow CO_2 + H_2)$ is an exothermic reaction and does not require such high temperatures to obtain a higher conversion of the CO as it is governed by equilibrium $(K_c = P_{CO_2}P_{H_2}/P_{CO}P_{H_2O})$. HyPr-RING is an acronym for hydrogen production by reaction-integrated novel gasification. HyPr-RINGs first proposed in 1998; and only after 2 years, Japan started this project to develop at a commercial scale HyPr-RING performs both conventional steam coal gasification and water-gas shift reaction in a single step in presence of CO_2 sorbent (i.e., CaO) [1]. Figure 4.1 illustrates a simple schematic for gasifier reactor of HyPr-RING process.

The raw materials supplied are hydrocarbons, CaO, and water; the major products are hydrogen and pure CO_2. HyPr-RING process involves two reactors: a gasifier and a regenerator. The HyPr-RING method concurrently performs coal gasification, CO_2 separation, and water–gas shift reaction in a single gasifier. This method is performed without combustion and produces very high concentration of hydrogen. The inclusion of CO_2 sorbent (i.e., CaO) to the reaction system serves dual purpose of CO_2 capture and hydrogen generation. Besides H_2S absorption $(CaO + H_2S \rightarrow CaS + H_2O)$, CaO can also catalyze NH_3 and tar decomposition. The spent CaO can be regenerated in another reactor called regenerator. Equation (4.1) illustrates the overall HyPr-RING process.

$$C + CaO + 2H_2O \rightarrow CaCO_3 + 2H_2 \Delta H^{\circ}_{298\ K} = -88kJ/mol \qquad (4.1)$$

CaO first reacts exothermally with high pressure steam to form reactive $Ca(OH)_2$. Simultaneously, coal reacts with steam to generate CO_2 and H_2. Further, $Ca(OH)_2$ and CaO lock CO_2 in the form of $CaCO_3$. In the regeneration reactor, $CaCO_3$ is

Fig. 4.1 Schematic of HyPr-RING gasifier process

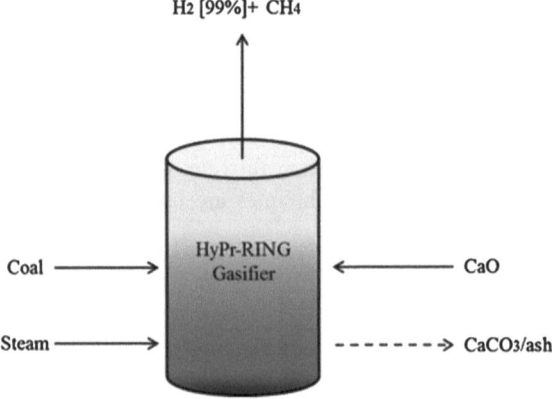

decomposed to regenerate CaO and liberate pure CO_2 gas stream ($CaCO_3 \rightarrow$ $CaO + CO_2$). In such way, the gasification-regeneration gets complete.

Previously, the successful integration of endothermic steam coal gasification and exothermic water–gas shift reaction has been experimentally demonstrated [2]. Shiying et al. performed a detailed thermodynamic study of the system between CaO, $Ca(OH)_2$, and $CaCO_3$ and reported that at 650 °C, CaO, and $Ca(OH)_2$ are in equilibrium at a steam partial pressure of 8.2 bar (the equilibrium favors $Ca(OH)_2$ when the pressure is >8.2 bar) [3]. Moreover, they also suggested that with the increase in CO_2 partial pressure, $CaCO_3$ can be formed from both CaO and $Ca(OH)_2$ phases. The phase equilibrium calculation also indicates that at any given temperature, higher CO_2 partial pressure is required for the formation of $CaCO_3$ using $Ca(OH)_2$ than CaO. They supported their thermodynamic calculation by performing experiments. Their experiments show that even at a low temperature (650 °C), calcium compounds were crystallized and formed bigger particles with a carbon conversion of 60–80 %. However, at high temperature (700 °C), solids melted and blocked the reactor. Moreover, the mass and energy flow was also calculated for a typical HyPr-RING process of 1,000 t coal per day [4]. These calculations showed that a fuel gas comprising an equilibrium mixture of 91 % H_2 and 9 % CH_4 can be obtained by gasification of coal at 650 °C and 30 bar. The amount of the fuel gas produced was calculated to be equivalent to 1.4 Nm^3/kg-coal, giving high cold-gas efficiency (=heat value of fuel gas/heat value of coal) of 0.77. The result shows that the equilibrium gas compositions are very sensitive to the temperature and pressure change. In particular, the yield of hydrogen doubles with an increase in temperatures from 650 to 700 °C. Moreover, the yield of hydrogen increases by 1.5 times with an increase in total pressure from 10 to 60 atm and in the steam partial pressure from 7 to 42 atm [5].

4.2.2 ZECA

Los Alamos National Laboratory proposed a new concept for hydrogen production called "zero emission coal alliance" (ZECA) [6]. ZECA integrates coal gasification without combustion, hydrogen production with CO_2 absorption, and regeneration. Interestingly, in ZECA process, coal is hydrogasified with hydrogen to produce methane. Figure 4.2 shows the schematic of ZECA process. As mentioned, this process involves three reactions and their operation conditions are mentioned in parentheses:

1. Coal is hydrogasified to produce methane

$$C + 2H_2 \rightarrow CH_4 (815 \,°C, 62 \text{ bars}) \tag{4.2}$$

2. Methane is reformed by steam to produce CO_2 and H_2, CO_2 is fixed by CaO to $CaCO_3$

Fig. 4.2 Schematic of ZECA process

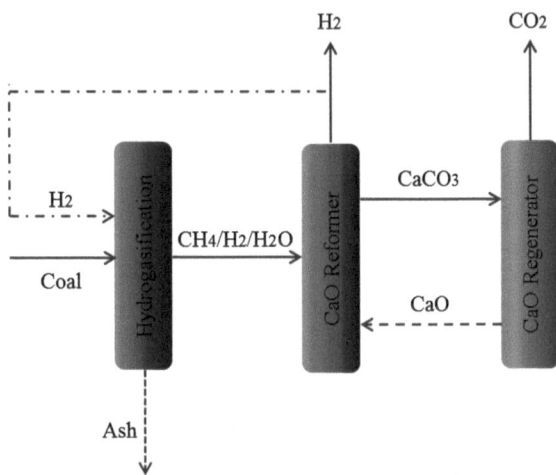

$$CH_4 + 2H_2O \rightarrow CO_2 + 4H_2(815\ °C,\ 30\ bars) \qquad (4.3)$$

3. $CaCO_3$ is sent to regenerator

$$CO_2 + CaO \leftrightarrow CaCO_3(1{,}130\ °C,\ \sim 1\ bar) \qquad (4.4)$$

As can be seen from Fig. 4.2, coal is first hydrogasified to produce CH_4. The hydrogasification reactor functions at 815 °C, 62 bars. The gas leaving this reactor is mainly composed of CH_4 and H_2 as well as steam, CO and CO_2. The produced gas from hydrogasification reactor can then be sent to a reformer with CaO bed material. In this reactor, CH_4 reacts with H_2O in presence of CaO to form H_2 and $CaCO_3$. The reformer reactor works at 815 °C, 30 bars. The used CO_2 sorbent as $CaCO_3$ can further be sent for CaO regeneration to a regenerator reactor, which operates at a very high temperature (1,130 °C) but ambient pressure.

4.2.3 CCR (Carbonation–Calcination Reaction)

The Ohio State University proposed a method to integrate coal gasification with a carbonation/calcination reactions (CCR) process to generate clean hydrogen [5]. Coal gasification produces the synthesis gas containing CO, CO_2, H_2, and H_2O (Fig. 4.3). The synthesis gas is then fed to water–gas shift reactor in the presence of CaO-based sorbent. The obtained $CaCO_3$ produced can further be sent to a rotary calciner and reused.

All the involved reactions in the integrated coal gasification and CCR process are mentioned below:

Fig. 4.3 Diagram of CCR process

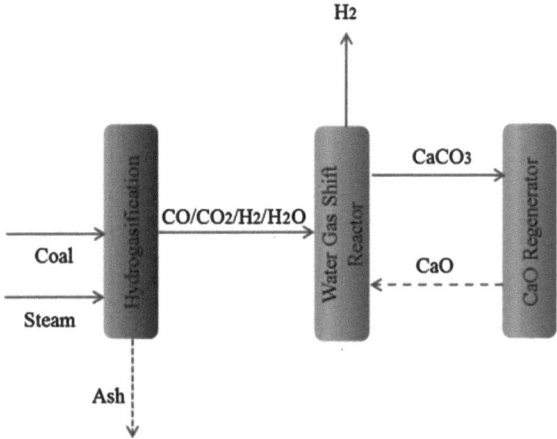

$$C + 2H_2O \rightarrow CO + 2H_2 \qquad (4.5)$$

$$CO + H_2O \rightarrow CO_2 + H_2 \qquad (4.6)$$

$$CO_2 + CaO \leftrightarrow CaCO_3 \qquad (4.7)$$

4.2.4 AGC: Advanced Gasification Combustion

The National Energy Technology Laboratory (NETL) and General Electric and Environmental Research Corporation (GE-EER) developed advanced gasification combustion (AGC) project. As shown in Fig. 4.4, AGC uses three fluidized bed reactors. In gasification reactor, coal is partly gasified with steam in the presence of CaO. CaO captures CO_2 and forms Ca-carbonates. Further, the CO_2 sorbent, CaO, is regenerated using the heat released by the combustion of char, which finally releases CO_2. Currently, AGC is performed at a pilot scale.

1. Coal is partly gasified with steam in the presence of CaO

$$C + H_2O \rightarrow CO + H_2 \qquad (4.8)$$

$$CO + H_2O \rightarrow CO_2 + H_2 \qquad (4.9)$$

2. Heat released by the combustion of char is used to regenerate CO_2 sorbent (CaO)

$$CO_2 + CaO \leftrightarrow CaCO_3 \qquad (4.10)$$

Fig. 4.4 Schematic of AGC process

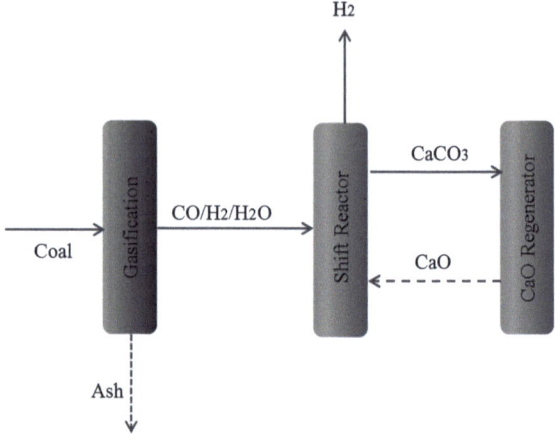

As can be seen, all the above-proposed techniques use CaO as a CO_2 sorbent. However, these methods are still in their development stage. The CO_2 capture concept is based on the earlier mentioned carbonation–calcination reaction cycle $[CaO(s) + CO_2(g) \leftrightarrow CaCO_3(s)]$. Figure 4.5 illustrates the equilibrium in the $C–H_2O–CaO$ system at 1 bar. The thermodynamic equilibrium calculation shows that the system can (i) produce hydrogen at a relatively lower temperature; (ii) capture CO_2 in the form of $CaCO_3$, which is stable in the reaction condition; and (iii) form $Ca(OH)_2$ as an intermediate, which decomposes at temperatures higher than 500 °C. The phase diagram suggests that at constant P_{CO_2}, stability of $CaCO_3$ decreases with decrease in P_{H_2O} [3]. The equilibrium of $C–H_2O–CaO$ system at various pressure and temperature is provided elsewhere [1]. It shows that in the $C–H_2O–CaO$ system, [CO], [CO_2], and [CH_4] are lower, and [H_2] is higher than

Fig. 4.5 Equilibrium in the system $C–H_2O–CaO$

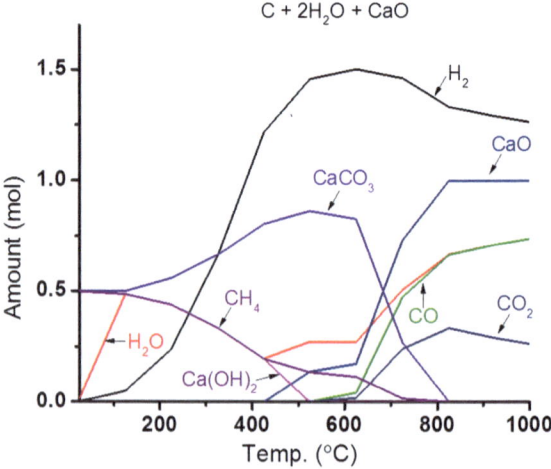

those in the C–H_2O system. [CO] and [CO_2] decreases with increasing pressure and reaches approximately zero at pressure >40 bars. The fixation of CO_2 by CaO is significant to reduce the [CO] and [CH_4] and increase [H_2].

4.3 Coal Gasification Using Non-fossil Energy Sources

4.3.1 Biomass

Co-gasification of coal and biomass offers several advantages over coal or biomass gasification [7]. Coal gasification emits high greenhouse gas compared to biomass gasification due to the high carbon content and low volatile percentage in coal [8, 9]. Liu et al. [8] reported that greenhouse gas emission index (GHGI) for coal gasification was 1.91 which significantly reduced to 0.96 when a mixture of 40 % biomass and 60 % coal was used as feedstock. Besides the obvious benefits of using renewable resources, it also allows biomass gasification to be performed at a larger scale with high efficiency and lower specific operating costs than conventional one [10, 11]. One such example is of IGCC plant at Buggenum (Netherlands), where co-gasification of up to 50 % w/w biomass were performed to reduce CO_2 emission [12]. In the same context, Maxim et al. [13] designed the concept of integrating CCS at IGCC plant based on coal gasification of coal and biomass.

Numerous laboratory-scale studies have been reported based on co-gasification of coal and biomass in fluidized and fixed bed reactors under varying conditions. It has been observed that synergy effects in co-pyrolysis and co-gasification lead to high reactivity [14], high fuel conversion [15], change in H_2/CO ratio in product gas [16], and reduced tar production [17]. Recently, Howaniec et al. [18] performed steam coal gasification of Polish hard coal and Salix Viminalis blends in a fixed bed reactor under ambient pressure and temperature of 700–900 °C. They concluded that in co-gasification of blends of 20 and 40 % w/w content of biomass, the total volume of product gas was increased when compared to coal or biomass gasification. However, at higher biomass content of 60 and 80 % w/w, a slight decrease in the volume of product gas was observed. In their recent study, a correlation between ash composition and the synergy effects was mentioned. Further, they also determined the optimal operating conditions of energy crop types and steam co-gasification in terms of hydrogen-rich gas production [19]. Moreover, Wang et al. [20] simulated a model comprises of two sub-models (i.e., combustion and gasification models) to test a single fluidized bed performing two-step gasification process and predicted the H_2 generation at various conditions. They observed that the molar concentration of hydrogen increases with the increase of steam/coke ratio, a value of 1.8 is highly desired. Moreover, a ratio of ¼ for coal/biomass can produce highest amount of hydrogen.

It is noteworthy that addition of biomass to coal gasification can not only mitigate carbon emission but also control the emission of sulfur and ash contained in coal. This could be achieved due to the biomass, which has almost no sulfur and low ash content [9]. However, biomass is expensive to gasify and produces large amount of tar. In Chap. 2, a summary of the increasing use of alkali to control the undesired tar formation during biomass gasification is provided. In essence, co-gasification can not only reduce the cost of feedstock but also control the problem of tar formation.

4.3.2 Solar-Driven Gasification

As gasification is an energy intensive process, it requires an external energy source to maintain a high-temperature environment inside the gasifier. Solar-driven gasification uses concentrated solar energy to provide such enormous energy and thus have potential to reduce overall CO_2 emission. The credit goes to Gregg et al. [21] who first demonstrated the solar-gasification of sub-bituminous coal, activated carbon, coke, and a mixture of coal and biomass in a fixed bed using a 23-kW solar furnace. The sunlight was focused on the bed through quartz window. Similarly, Taylor et al. [22] used a 2-kW vertical-beam solar furnace to gasify carbonaceous materials in a packed bed reactor. They compared the performance of fluidized bed reactor to a packed bed reactor and found that the fraction of the incident solar energy utilized to produce CO (stored) was 10 % in the case of the fluidized bed reactor and 40 % for the packed bed reactor. Another study dealt with a bench scale unit of the UNH Gas Recirculation System for coal carbonization that has been tested using an electric heater to simulate solar energy with nitrogen as the heat carrier gas [23]. This study shows the successful carbonization of coal by indirectly transferring simulated solar heat, using the gas recirculation technique. The products obtained are similar to those from a conventional carbonization unit. To date, significant improvements have been made in this field. However, the main challenge lies in scaling up the solar-based methods. Recently, Dincer et al. [24] presented a detail study on the role of solar energy for hydrogen production and could be worth reading.

4.4 In Situ CO_2 Capture Using NaOH

The inclusion of sodium hydroxide to the steam coal gasification technology transforms the process to a much simplified-single step reaction (4.11).

$$2NaOH(s) + C(s) + H_2O(g) \rightarrow Na_2CO_3(s) + 2H_2(g)$$
$$\Delta H_{327\,°C} = 64.58 \text{ kJ/mol} \tag{4.11}$$

A brief introduction to the method is given in Sect. 2.1.1. As mentioned earlier, sodium hydroxide can facilitate steam coal gasification process (4.12) to generate hydrogen at a relatively low temperature.

$$2C(s) + 3H_2O(g) \rightarrow CO(g) + CO_2(g) + 3H_2(g)$$
$$\Delta H_{327\,°C} = 95.73 \text{ kJ/mol} \tag{4.12}$$

Figure 4.6 shows the thermodynamic calculation for sodium hydroxide-assisted coal gasification process $[NaOH(s) + C(s) + H_2O(g)]$ using Factsage^{TM} software. The use of sodium hydroxide to the coal-steam system requires less energy input (95.73 kJ/mol reduced to 64.58 kJ/mol at 327 °C) [19]. The system not only locks CO_2 in the form of soda ash but also generates hydrogen. Moreover, the system does not produce complex mixture of gases.

Table 4.1 compares the energy requirement for sodium hydroxide-assisted coal gasification process with the conventional coal gasification method. It is apparent that the inclusion of alkali has potential to lower the operating temperature and carbon dioxide emission as well. Consequently, relatively less amount of coal would be required to operate such modified reaction. The by-product of reaction (4.11), sodium carbonate (Na_2CO_3), has huge application in different chemical sectors such as glass manufacturing, electrolyte, textiles, and domestic use.

Sodium hydroxide has previously been used for hydrogen production in industries. For instance, the black liquor gasification process uses alkali hydroxide for both hydrogen production and carbon sequestration. In a typical pulping process for paper production, approximately one-half of the raw materials are transformed to pulp and other half is dissolved in the black liquor. The black liquor solution has well-dispersed carbonaceous material, steam, and alkali metal which act as energy source for the plant. Reactions (4.13) and (4.14) dominate in the presence of carbonaceous material and water in the liquor:

Fig. 4.6 Calculated equilibrium in the system $2NaOH(s) + C(s) + H_2O(g)$

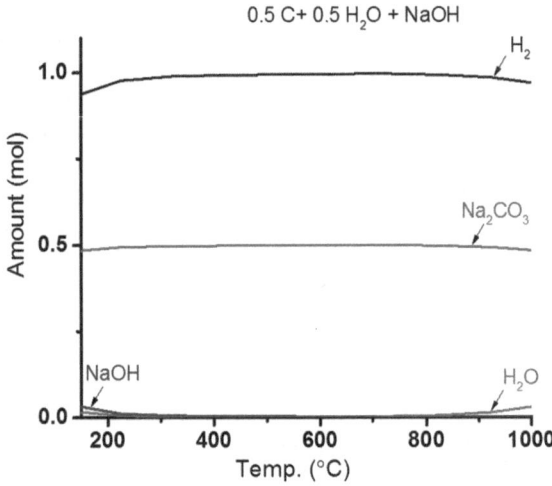

Table 4.1 Thermodynamic properties for different hydrogen production methods after inclusion of NaOH [20]

	$C + H_2O$	$NaOH + C + H_2O$
Temperature (°C)	800–1,200	500–700
Enthalpy (ΔH, kJ/mol)	95.73(327 °C)	64.58 (327 °C)
Mixture of product gases	CO, CO_2, H_2	H_2
Coal/H_2(g/g)	3.73	3.49
CO_2/H_2(g/g)	13.67	1.80

$$C(s) + H_2O(g) = CO(g) + H_2(g) \qquad (4.13)$$

$$CO(g) + H_2O(g) = CO_2(g) + H_2(g) \qquad (4.14)$$

However, reaction (4.14) is thermodynamically limited and thus never proceeds toward completion and that prohibits hydrogen concentration to exceed a certain limit. Interestingly, in the presence of NaOH, the equilibrium can be shifted to drive reaction (4.14) toward completion and maximum hydrogen concentration can be achieved. As can be expected, the concentration of undesired CO and CO_2 is reduced considerably in the product gases.

We performed the experimental study over reaction (4.11) and determined the optimal catalysts and their sizes [21, 22]. Prior to the selection of suitable catalysts for the modified coal gasification reaction, we established the correlation of coal particle size with the reaction yield. Coal particles were mechanically milled for 0, 1, 2, and 4 h. The corresponding average particle size was 15.07, 9.29, 6.27, and 13.97 µm for raw, 1, 2, and, 4 h, respectively. Further, reaction (4.11) was carried out at 600 °C using various coal particle sizes. The experimental results showed that higher coal particle size leads to a lower % conversion of sodium hydroxide to soda.

Moreover, catalytic study was also performed over 2-h-ball-milled coal (mean size of 6.27 µm) using nickel (100 mesh) and Raney nickel (−325 mesh) at 600 °C. The experimental results exhibited that in the presence of Raney nickel catalysts, reaction has the maximum yield at any time. Raney nickel possesses highly porous microstructure and thus is more active than nickel catalyst for reaction (3.8). After the selection of appropriate catalysts for the reaction, we performed mechanical milling to correlate the reaction yield with catalysts particle size. Therefore, Raney nickel catalysts were milled for 0, 1, 2, or 4 h time and the corresponding crystallite size (calculated using X-ray diffraction pattern and Scherrer equation) was calculated. The calculation showed that the crystallite sizes of raw 1-, 2-, or 4-h-ball-milled Raney nickel are 413, 347, 209, and 190 Å, respectively. As expected, the crystallite size decreased due to mechanical milling. Moreover, the average particle sizes of the variously milled Raney nickel catalysts were 11.74, 12.18, 15.72, and 13.72 µm for raw, 1-, 2-, and 4-h-milled particles. Two-hour-milled Raney nickel showed the highest reaction kinetics at 500 °C. The reason could be attributed to both geometrical and morphological factors. The geometrical factors include combined effect of the particle and crystallite size. As can be seen, mechanical milling was performed to correlate the catalytic performance of Raney nickel catalyst with its particle and crystallite size. The average particle size of the Raney nickel increased till 2 h milling time and that indicates the dominating cold welding process. However, prolonged milling reduced the average particle size for 4-h-milled sample and reveals the increasing dominance of fracture process. The reduced performance of 4-h-ball-milled Raney nickel can also be due to either a high mechanical abrasion of the catalysts surface or thermally induced deactivation. During longer milling hours, the temperature inside the jar increases that influences the surface area of particles. Moreover, 2-h-milled Raney nickel catalyst has maximum % reduction in the crystallite size. Therefore, a large number of generations and movement of dislocations or vacancies led to a high catalytic performance for 2-h-ball-milled Raney nickel catalysts.

As seen above, the inclusion of sodium hydroxide can reduce the operating condition of coal gasification process. The use of 2-h-milled Raney nickel (milled with powder to ball ratio of 1:54) can further lower this condition. However, there are other important factors (such as corrosiveness and sources of sodium hydroxide, recyclability of catalysts, and reactor design) which must be investigated in detail to scale up the reaction. As indicated in the previous chapters, the inclusion of sodium hydroxide to fossil-fuel-based hydrogen production methods can have huge benefits if renewable sources can be the prime energy contributor.

4.5 Conclusion

It is evident that significant work has been dedicated to modify the conventional coal gasification process. Most of these modified routes primarily integrate CO$_2$ sorbents (CaO, NaOH) or renewable resources (biomass, solar) in the reaction system to reduce vast CO$_2$ emissions. However, at present, these new reaction paths also inherit challenges and needs to be resolved soon. Overcoming these challenges can help in establishing the hydrogen economy.

References

1. Lin S, Suzuki Y, Hatano H, Harada M (2002) Developing an innovative method, HyPr-RING, to produce hydrogen from hydrocarbons. Energy Convers Manage 43:1283–1290. doi:10. 1016/S0196-8904(02)00014-6
2. Lin S, Harada M, Suzuki Y, Hatano H (2004) Continuous experiment regarding hydrogen production by coal/CaO reaction with steam (I) gas products. Fuel 83:869–874. doi:10.1016/j. fuel.2003.10.030
3. Lin S, Harada M, Suzuki Y, Hatano H (2006) Continuous experiment regarding hydrogen production by Coal/CaO reaction with steam (II) solid formation. Fuel 85:1143–1150. doi:10. 1016/j.fuel.2005.05.029
4. Lin S, Harada M, Suzuki Y, Hatano H (2005) Process analysis for hydrogen production by reaction integrated novel gasification (HyPr-RING). Energy Convers Manage 46:869–880. doi:10.1016/j.enconman.2004.06.008
5. Fan LS (2011) Chemical looping systems for fossil energy conversions (Chap. 6). Wiley, NJ, 382 p

6. Gupta R (2008) Hydrogen fuel: production, transport and storage (Chap. 3). CRC Press, FL, pp 114–124

7. Cormos CC (2012) Hydrogen and power co-generation based on coal and biomass/solid wastes co-gasification with carbon capture and storage. Int J Hyd Energy 37:5637–5648. doi:10.1016/j.ijhydene.2011.12.132

8. Liu G, Larson ED, Williams RH, Kreutz TG, Guo X (2011) Making fischer-tropsch fuels and electricity from coal and biomass: performance and cost analysis. Energy Fuels 25:415–437. doi:10.1021/ef101184e

9. Kumabe K, Hanaoka T, Fujimoto S, Minowa T, Sakanishi K (2007) Cogasification of woody biomass and coal with air and steam. Fuel 86:684–689. doi:10.1016/j.fuel.2006.08.026

10. Valero A, Uson S (2006) Oxy-co-gasification of coal and biomass in an integrated gasification combined cycle (IGCC) power plant. Energy 31:1643–1655. doi:10.1016/j.energy.2006.01.005

11. Chmielniak T, Sciazko M (2003) Co-gasification of biomass and coal for methanol synthesis. Appl Energy 74:393–403. doi:10.1016/S0306-2619(02)00184-8

12. Prins M, Ptasinski K, Janssen F (2007) From coal to biomass gasification: comparison of thermodynamic efficiency. Energy 32:1248–1259. doi:10.1016/j.energy.2006.07.017

13. Maxim V, Cormos C, Agachi P (2011) Design of integrated gasification combined cycle plant with carbon capture and storage based on co-gasification of coal and biomass. Comput Aided Chem Eng 29:1904–1908. doi:10.1016/B978-0-444-54298-4.50159-8

14. Collot A, Zhuo Y, Dugwell D, Kandiyoti R (1999) Co-pyrolysis and co-gasification of coal and biomass in bench-scale fixed-bed and fluidised bed reactors. Fuel 78:667–679. doi:10.1016/S0016-2361(98)00202-6

15. Lapuerta M, Hernández J, Pazo A, López J (2008) Gasification and co-gasification of biomass wastes: effect of the biomass origin and the gasifier operating conditions. Fuel Process Tech 89:828–837. doi:10.1016/j.fuproc.2008.02.001

16. Li K, Zhang R, Bi J (2010) Experimental study on syngas production by co-gasification of coal and biomass in a fluidized bed. Int J Hyd Energy 35:2722–2726. doi:10.1016/j.ijhydene.2009.04.046

17. Sjöström K, Chen G, Yu Q, Brage C, Rosén C (1999) Promoted reactivity of char in co-gasification of biomass and coal: synergies in the thermochemical process. Fuel 78:1189–1194. doi:10.1016/S0016-2361(99)00032-0

18. Howaniec N, Smoliński A, Stańczyk K, Pichlak M (2011) Steam co-gasification of coal and biomass derived chars with synergy effect as an innovative way of hydrogen-rich gas production. Int J Hydrogen Energy 36:14455–14463. doi:10.1016/j.ijhydene.2011.08.017

19. Howaniec N, Smoliński A (2014) Effect of fuel blend composition on the efficiency of hydrogen-rich gas production in co-gasification of coal and biomass. Fuel 128:442–450. doi:10.1016/j.fuel.2014.03.036

20. Wang L, Dun Y, Xiang X, Jiao Z, Zhang T (2011) Thermodynamics research on hydrogen production from biomass and coal co-gasification with catalyst. Int J Hydrogen Energy 36:11676–11683. doi:10.1016/j.ijhydene.2011.06.064

21. Gregg DW, Taylor RW, Campbell JH, Taylor JR, Cotton A (1980) Solar gasification of coal, activated carbon, coke and coal and biomass mixtures. Sol Energy 25:353–364. doi:10.1016/0038-092X(80)90347-3

22. Taylor RW, Berjoan R, Coutures JP (1983) Solar gasification of carbonaceous materials. Sol Energy 30:513–525. doi:10.1016/0038-092X(83)90063-4

23. Mathur VK, Breault RW, Lakshmanan S (1983) Coal gasification using solar energy. Sol Energy 30:433–440. doi:10.1016/0038-092X(83)90113-5

24. Dincer I, Joshi A (2013) Solar based hydrogen production systems. Springer, New York. doi:10.1007/978-1-4614-7431-9